Die Grundlehren der mathematischen Wissenschaften

in Einzeldarstellungen
mit besonderer Berücksichtigung
der Anwendungsgebiete

Band 207

William F. Donoghue, Jr.

Monotone Matrix Functions and Analytic Continuation

Springer-Verlag
Berlin Heidelberg New York 1974

William F. Donoghue, Jr.
University of California, Irvine, California 92664, U.S.A.

AMS Subject Classifications (1970):
Primary 26 A 48, 30 A 14, 47 A 10
Secondary 15 A 18, 30 A 31, 30 A 32, 30 A 80

ISBN 3-540-06543-1 Springer-Verlag Berlin Heidelberg New York
ISBN 0-387-06543-1 Springer-Verlag New York Heidelberg Berlin

 Library of Congress Catalog Card Number 73-15293. Printed in Germany. Monophoto typesetting and offset printing: Zechnersche Buchdruckerei, Speyer. Bookbinding: Konrad Triltsch, Würzburg.

Foreword

A Pick function is a function that is analytic in the upper half-plane with positive imaginary part. In the first part of this book we try to give a readable account of this class of functions as well as one of the standard proofs of the spectral theorem based on properties of this class. In the remainder of the book we treat a closely related topic: Loewner's theory of monotone matrix functions and his analytic continuation theorem which guarantees that a real function on an interval of the real axis which is a monotone matrix function of arbitrarily high order is the restriction to that interval of a Pick function. In recent years this theorem has been complemented by the Loewner-FitzGerald theorem, giving necessary and sufficient conditions that the continuation provided by Loewner's theorem be univalent.

In order that our presentation should be as complete and transparent as possible, we have adjoined short chapters containing the information about reproducing kernels, almost positive matrices and certain classes of conformal mappings needed for our proofs.

We may say that the book is almost elementary. The reader is supposed only to have a reasonable acquaintance with analytic function theory and some Hilbert space theory. The book can be read by most first year graduate students. The reader who is primarily interested in the Loewner theory may well skip chapters IV and V on the Fatou theorems and the spectral theorem, since the results of those chapters are never invoked later in the book. Nevertheless, a book on Pick functions which omitted those topics would be incomplete indeed.

The greater part of the text was written when the author, on sabbatical leave from the University of California, enjoyed the large hospitality of the Mathematics Institute of the University of Lund. The influence of N. Aronszajn pervades the book, although his name never appears in the text. We are indebted to L. Gårding for valuable literary advice.

Most of the book follows the thought of the late Charles Loewner, one of the great mathematicians of the age. If our account in any way reflects his unpretentious and honest approach to mathematics, it will be all the success that the author desires.

Contents

I. Preliminaries 1
II. Pick Functions 18
III. Pick Matrices and Loewner Determinants 34
IV. Fatou Theorems 42
V. The Spectral Theorem 50
VI. One-Dimensional Perturbations 63
VII. Monotone Matrix Functions 67
VIII. Sufficient Conditions 78
IX. Loewner's Theorem 85
X. Reproducing Kernels 88
XI. Nagy-Koranyi Proof of Loewner's Theorem 94
XII. The Cauchy Interpolation Problem 100
XIII. Interpolation by Pick Functions 117
XIV. The Interpolation of Monotone Matrix Functions 128
XV. Almost Positive Matrices 134
XVI. The Analytic Continuation of Bergman Kernels 140
XVII. The Loewner-FitzGerald Theorem 146
XVIII. Loewner's Differential Equation 154
XIX. More Analytic Continuation 165

Notes and Comment 170

Bibliography 176

Index 181

Chapter I. Preliminaries

1. The Degree of a Rational Function

Let the rational function $f(z)$ be represented as the ratio of two relatively prime polynomials: $f(z)=p(z)/q(z)$ where d' is the degree of $p(z)$ and d'' the degree of $q(z)$. We define the degree of the rational $f(z)$ as $d=\max(d',d'')$. It can be shown that this degree is the same as the Brouwer degree obtained when $f(z)$ is regarded as a continuous mapping of the Riemann sphere into itself.

For any value of the constant λ the number of roots of the equation $f(z)=\lambda$ is exactly d, the degree of $f(z)$. Here, of course, multiple roots must be counted as often as their multiplicity requires, and a possible root at infinity must be counted if it occurs. For all but finitely many values of λ, however, these roots will all be finite and simple.

It is easy to show that if $f(z)$ has degree d' and $g(z)$ has degree d'', then both their sum and their product have degrees at most $d'+d''$. Again, the degree of the derivative $f'(z)$ is at most twice that of $f(z)$.

If d is the degree of the rational $f(z)$ and $f(z)$ vanishes for $d+1$ points, then $f(z)=0$ identically. From this circumstance it follows that a rational function of degree d is completely determined by its behaviour at $2d+1$ points. For if two functions of degree d coincide at $2d+1$ points their difference has at least $2d+1$ zeros and is of degree at most $2d$.

2. Divided Differences

For a complex-valued function $f(x)$ defined on a subset of the complex plane we introduce the divided differences as follows: Let $\lambda_1, \lambda_2, \lambda_3, \ldots$ be a sequence of distinct points; we set

$$[\lambda_1]=f(\lambda_1),$$

$$[\lambda_1,\lambda_2]=\frac{f(\lambda_1)-f(\lambda_2)}{\lambda_1-\lambda_2}, \quad \text{and inductively}$$

$$[\lambda_1,\lambda_2,\ldots,\lambda_{k+1}]=\frac{[\lambda_1,\lambda_2,\ldots,\lambda_k]-[\lambda_2,\lambda_3,\ldots,\lambda_{k+1}]}{\lambda_1-\lambda_{k+1}}.$$

For example, if c is a constant which is different from any of the λ_i, we easily verify when $f(x)=1/(c-x)$ that $[\lambda_1,\lambda_2]=1/((c-\lambda_1)(c-\lambda_2))$ and more generally, that $[\lambda_1,\lambda_2,\ldots,\lambda_n]=1\Big/\prod_{i=1}^{n}(c-\lambda_i)$. If we suppose that the function f is analytic in a region of the complex plane containing a rectifiable curve C which surrounds the points λ_i then, from Cauchy's Integral formula, when x is within C

$$f(x)=\frac{1}{2\pi i}\int_C \frac{f(\zeta)}{\zeta-x}\,d\zeta$$

and accordingly

$$[\lambda_1,\lambda_2,\ldots,\lambda_n]=\frac{1}{2\pi i}\int_C \frac{1}{\prod_{k=1}^{n}(\zeta-\lambda_k)}\,f(\zeta)\,d\zeta$$

and from the calculus of residues we finally obtain

$$[\lambda_1,\lambda_2,\ldots,\lambda_n]=\sum_{k=1}^{n}\frac{f(\lambda_k)}{\prod_{j\neq k}(\lambda_k-\lambda_j)}.$$

Since the formula above satisfies the inductive rule defining the divided difference it is evidently an algebraic identity and does not depend on the hypothesis that $f(z)$ is analytic, or, in fact, regular in any way. It is also evident that the divided differences are symmetric functions of their arguments, hence independent of the order in which the λ's are taken.

It is often convenient to compute with divided differences without the hypothesis that the points λ_i are distinct. It is then necessary to suppose that the function $f(z)$ is sufficiently smooth, so that the difference becomes a continuous function of the variables λ_i. In the case when $f(z)$ is analytic in an appropriate region, the integral formula which we have obtained is still valid, since it is continuous in the λ_i. Suppose we consider l distinct points $x_1, x_2, x_3, \ldots, x_l$ and seek an expression for the divided difference

$$[x_1, x_1, \ldots, x_2, x_2, \ldots, x_l, x_l, \ldots, x_l]$$

where the point x_1 occurs $1+v_1$ times, the point x_2 occurs $1+v_2$ times etc. The difference then involves $l+\sum v_i$ entries. From the integral formula we have

$$[x_1, x_1, \ldots, x_l]=\frac{1}{2\pi i}\int_C \frac{f(\zeta)\,d\zeta}{P(\zeta)}$$

where the polynomial $P(\zeta)$ is $\prod_{k=1}^{l} (\zeta - x_k)^{1+\nu_k}$. The integral can be computed by residues as before, and we obtain

$$\sum_{k=1}^{l} \frac{1}{\nu_k!} \left(\frac{d}{d\zeta}\right)^{\nu_k} \left(\frac{f(\zeta)(\zeta - x_k)^{1+\nu_k}}{P(\zeta)}\right)\Bigg|_{\zeta = x_k}.$$

We will rarely need these explicit formulas in the sequel, but note merely that the divided difference depends on the values of the function and its derivatives up to the order ν_k at x_k for all k. This formula is valid for all functions having sufficiently many continuous derivatives.

In our notation for a divided difference it will occasionally be necessary to indicate the function; in this case the function will occur as a subscript, and the differences which we have discussed so far would be written $[\lambda_1, \lambda_2, \ldots, \lambda_n]_f$. It is obvious that the divided difference is a linear operator:

$$[\lambda_1, \lambda_2, \ldots, \lambda_n]_{af+bg} = a[\lambda_1, \lambda_2, \ldots, \lambda_n]_f + b[\lambda_1, \lambda_2, \ldots, \lambda_n]_g$$

and we have tacitly made use of this fact when we brought the difference operator under the integral sign in our previous argument.

Let $\{\lambda_i\}$ be an infinite sequence of not necessarily distinct points. Associated with this sequence is a set of Newtonian interpolation polynomials defined as follows:

$$P_0(z) = 1, \quad P_1(z) = z - \lambda_1, \quad P_2(z) = (z - \lambda_1)(z - \lambda_2), \ldots, P_k(z) = \prod_{i=1}^{k} (z - \lambda_i).$$

We next compute the quantity

$$[\lambda_1, \lambda_2, \ldots, \lambda_n]_{P_k} = \frac{1}{2\pi i} \int_C \frac{P_k(\zeta)}{P_n(\zeta)} d\zeta$$

where the curve C is a circle containing all the λ_i in its interior. If $n \leqq k$, P_n divides P_k and the integrand is analytic. It follows that the divided difference in that case is 0. If $n = k+1$ the numerator divides the denominator and the integrand becomes $(\zeta - \lambda_n)^{-1}$. Accordingly

$$[\lambda_1, \lambda_2, \ldots, \lambda_{k+1}]_{P_k} = \frac{1}{2\pi i} \int_C \frac{d\zeta}{\zeta - \lambda_{k+1}} = 1.$$

Finally, if $n > k+1$, the absolute value of the integrand diminishes with increasing $|\zeta|$ at least as rapidly as $|\zeta|^{-2}$, and it follows that the integral vanishes, since we can choose the circle C arbitrarily large. We find, therefore, that

$$\begin{aligned} [\lambda_1, \lambda_2, \ldots, \lambda_n]_{P_k} &= 1 \quad \text{if } n = k+1 \\ &= 0 \quad \text{otherwise}. \end{aligned}$$

The system of polynomials $P_k(z)$ obviously forms a basis for the vector space of polynomials in z: they are linearly independent and there is exactly one polynomial of each degree, since the degree of $P_k(z)$ is k. Accordingly, an arbitrary polynomial $f(z)$ admits a unique expansion of the form

$$f(z)=\sum_{k=0}^{N} A_k P_k(z).$$

Using the linearity of the divided difference operator we have

$$[\lambda_1, \lambda_2, \ldots, \lambda_n]_f=\sum_{k=0}^{N} A_k[\lambda_1, \lambda_2, \ldots, \lambda_n]_{P_k}=A_{n-1}.$$

Thus

$$f(z)=\sum_{k=0}^{N} [\lambda_1, \lambda_2, \ldots, \lambda_{k+1}]_f P_k(z).$$

When the points λ_i all coincide, this reduces to the usual Taylor expansion of $f(z)$ about that point. Our argument also shows that if $f(z)$ is a polynomial of degree N, the divided difference $[\lambda_1, \lambda_2, \ldots, \lambda_n]_f=0$ if $n \geqq N+2$.

We now make use of the Newtonian interpolation polynomials to solve the following simple interpolation problem. We suppose that l distinct points

$$x_1, x_2, \ldots, x_l$$

are given, as well as equally many non-negative integers

$$\nu_1, \nu_2, \ldots, \nu_l$$

and it is required to find a polynomial which, together with its derivatives of order $\leqq \nu_k$, has prescribed values at each x_k. It is clear that the data of the problem include the value of the divided difference $[\lambda_1, \lambda_2, \ldots, \lambda_n]$ whenever the set of numbers inside the brackets is a subset of the set

$$S=\{x_1, x_1, \ldots, x_1, x_2, \ldots, x_l\}$$

formed by taking each x_k $1+\nu_k$ times. The cardinal of S is $N=l+\sum \nu_k$. To obtain a solution to the interpolation problem, we have only to rewrite S in the form $\{\lambda_1, \lambda_2, \ldots, \lambda_N\}$ and to take the corresponding Newtonian interpolation polynomials. The coefficients of the expansion of the solution in terms of these polynomials are known.

Perhaps a word of warning is appropriate here. The set S of N numbers has 2^N-1 non-empty subsets and so there are equally many divided differences $[\lambda_1, \lambda_2, \ldots, \lambda_n]$ to be considered. These are certainly not independent quantities. Indeed, if all $\nu_k=0$, the values of the l

differences $[x_k]$ completely determine the higher differences in view of the inductive definition of those differences. Thus, whenever an interpolation problem is considered, and we shall later treat some in this book, it must be tacitly supposed that the data of the problem are consistent. We cannot ask for a function vanishing at $z=0$ and $z=1$ for which the divided difference $[0,1]$ is 7.

We make use of this type of interpolation by polynomials to solve the following problem. Let $u(z)$ and $v(z)$ be two polynomials and $f(z)$ their product; we suppose the numbers $\lambda_1, \lambda_2, \ldots, \lambda_n$ given and seek an expression for $[\lambda_1, \lambda_2, \ldots, \lambda_n]_f$ in terms of the divided differences of $u(z)$ and $v(z)$.

Take first the Newtonian interpolation polynomials $P_k(z)$ associated with the sequence λ_i and write

$$u(z) = \sum_{k=0}^{N} A_k P_k(z).$$

If the degree of $u(z)$ is greater than n, it is necessary, of course, to adjoin further points to the sequence. In a somewhat similar way we expand $v(z)$, only this time we re-order the sequence, writing

$$\lambda_n, \lambda_{n-1}, \ldots, \lambda_1$$

and making use of the Newtonian interpolation polynomials $Q_j(z)$ defined as follows. $Q_0(z)=1$, $Q_1(z)=z-\lambda_n$, $Q_2(z)=(z-\lambda_n)(z-\lambda_{n-1})$ etc. Again, if necessary, we adjoin further points to the sequence to obtain

$$v(z) = \sum_{j=0}^{M} B_j Q_j(z).$$

Accordingly

$$f(z) = \sum_{k=0}^{N} \sum_{j=0}^{M} A_k B_j P_k(z) Q_j(z).$$

The polynomial $P_k(z)Q_j(z)$ is of degree $k+j$, so if $k+j \leqq n-2$ the divided difference $[\lambda_1, \lambda_2, \ldots, \lambda_n]_{P_k Q_j}$ vanishes. On the other hand, if $k+j \geqq n$ that divided difference may be written in integral form as

$$\frac{1}{2\pi i} \int_C \frac{P_k(\zeta) Q_j(\zeta) d\zeta}{(\zeta-\lambda_1)(\zeta-\lambda_2)\cdots(\zeta-\lambda_n)}$$

where C is an appropriate circle. However, the denominator in the integrand divides the numerator, so the integral vanishes. It follows that only values of k and j such that $k+j=n-1$ contribute to the divided differences; in this case $[\lambda_1, \lambda_2, \ldots, \lambda_n]_{P_k Q_j} = (1/2\pi i) \int_C dz/(z-\lambda_{k+1}) = 1$ and

$$[\lambda_1, \lambda_2, \ldots, \lambda_n]_f = \sum_{k=0}^{n-1} A_k B_{n-1-k}$$
$$= \sum_{k=0}^{n-1} [\lambda_1, \lambda_2, \ldots, \lambda_{k+1}]_u [\lambda_{k+1}, \lambda_{k+2}, \ldots, \lambda_n]_v .$$

We may call this the Leibnitz rule for the divided differences, since this reduces to the usual Leibnitz formula for the differentiation of a product when the points λ_i all coincide.

In this book we will compute with divided differences only for two possible cases: either the functions will be analytic in a region containing the interpolation points λ_i, and so the integral formulas which we have given are valid, or the function will be a real function on an interval of the real axis, subject to certain smoothness conditions. These conditions will be enough to ensure that the divided differences are continuous and symmetric functions of their arguments. The following lemma will be useful.

Lemma. *Let $f(x)$ be C^{n-1} on the interval $a<x<b$ and $\lambda_1, \lambda_2, \ldots, \lambda_n$ n points of that interval. Then there exists a point ξ in the smallest interval containing the λ's such that*

$$[\lambda_1, \lambda_2, \ldots, \lambda_n] = \frac{f^{(n-1)}(\xi)}{(n-1)!} .$$

Proof. We first assume that the points λ_i are distinct. Because the divided difference is symmetric, we are also at liberty to take $\lambda_1 = \min \lambda_i$ and $\lambda_n = \max \lambda_i$. Now we form the polynomial

$$p(x) = \sum_{k=0}^{n-1} [\lambda_1, \lambda_2, \ldots, \lambda_{k+1}]_f P_k(x)$$

which coincides with $f(x)$ at each point λ_i, since the divided differences computed for these points are the same for both f and p. The difference

$$F(x) = f(x) - p(x)$$

has n distinct zeros in the closed interval $\lambda_1 \leqq x \leqq \lambda_n$, and so, by Rolle's theorem, its derivative has $n-1$ such zeros in that interval. Continuing in this way, we infer that the derivative of order $n-1$ of $F(x)$ has at least one zero in the interval. Accordingly at some point ξ

$$F^{(n-1)}(\xi) = 0 = f^{(n-1)}(\xi) - [\lambda_1, \lambda_2, \ldots, \lambda_n]_f (n-1)!$$

When the λ_i are not distinct we make use of the fact that the divided difference is a continuous function of its arguments to find systems

$$[\lambda'_1, \lambda'_2, \ldots, \lambda'_n] = \frac{f^{(n-1)}(\xi')}{(n-1)!}$$

converging to $[\lambda_1, \lambda_2, \ldots, \lambda_n]$. Evidently the corresponding numbers ξ' have a limit point in the closed interval $\lambda_1 \leqq x \leqq \lambda_n$, and from the continuity of $f^{(n-1)}(x)$ the lemma follows.

There is also a Remainder theorem for the Newtonian interpolation.

Theorem. *Let $\lambda_1, \lambda_2, \ldots, \lambda_n$ be n distinct points and $R_{n-1}(x)$ be defined by the equation*

$$f(x) = \sum_{k=0}^{n-1} [\lambda_1, \lambda_2, \ldots, \lambda_{k+1}]_f P_k(x) + R_{n-1}(x).$$

Then

$$R_{n-1}(x) = [\lambda_1, \lambda_2, \ldots, \lambda_n, x]_f P_n(x).$$

Proof. We may write $f(\lambda_{n+1})$ in two different ways:

$$f(\lambda_{n+1}) = \sum_{k=0}^{n-1} [\lambda_1, \lambda_2, \ldots, \lambda_{k+1}]_f P_k(\lambda_{n+1}) + R_{n-1}(\lambda_{n+1})$$

and

$$f(\lambda_{n+1}) = \sum_{k=0}^{n} [\lambda_1, \lambda_2, \ldots, \lambda_{k+1}]_f P_k(\lambda_{n+1})$$

and deduce that

$$R_{n-1}(\lambda_{n+1}) = [\lambda_1, \lambda_2, \ldots, \lambda_n, \lambda_{n+1}]_f P_n(\lambda_{n+1}).$$

The numbers $\lambda_1, \lambda_2, \ldots, \lambda_n$ being given, λ_{n+1} can be arbitrary, whence

$$R_{n-1}(x) = [\lambda_1, \lambda_2, \ldots, \lambda_n, x]_f P_n(x)$$

as desired. It is clear that the same theorem is true if the points are not necessarily distinct, provided the function $f(x)$ is sufficiently smooth.

3. Positive Matrices

A square matrix $A = a_{ij}$ is called a positive matrix if the corresponding quadratic form

$$\sum \sum a_{ij} z_i \bar{z}_j$$

is non-negative for any choice of the complex numbers z_i. It is easy to see that in this case the diagonal elements of the matrix are $\geqq 0$ and that the matrix is symmetric, i.e. $a_{ij} = \bar{a}_{ji}$. If we select a complete orthonormal system e_i in the n-dimensional Hilbert space, the matrix gives rise in a known way to a linear transformation of that space determined by the equations

$$A u = \sum \xi_k A e_k = \sum \sum a_{ik} \xi_k e_i$$

when $u=\sum \xi_k e_k$. It is easy to see that we will have $a_{ij}=(Ae_j, e_i)$. The positiveness of the quadratic form means that for every vector u in the Hilbert space $(Au, u)\geqq 0$, and so, in particular, the eigenvalues of the transformation A are all non-negative. Since the product of these eigenvalues is the determinant of the matrix a_{ij}, we see that the determinant in question is also non-negative.

Let A be of order 2:

$$A=\begin{bmatrix} a_{11} & a_{12} \\ a_{21} & a_{22} \end{bmatrix} \quad \text{where } a_{12}=\bar{a}_{21} \quad \text{and the diagonal elements are real}.$$

Then the positiveness of A is equivalent to the positiveness of the trace $a_{11}+a_{22}=$ sum of the eigenvalues, and the positiveness of the determinant $a_{11}a_{22}-|a_{12}|^2=$ product of the eigenvalues. We may write the latter inequality

$$|(Ae_2, e_1)|^2 \leqq (Ae_1, e_1)(Ae_2, e_2)$$

and see that it is merely the Schwarz inequality for the positive quadratic form (Au,u) on the Hilbert space. It is clear that this equation holds whatever be the order of A:

$$|a_{ij}|^2 \leqq a_{ii} a_{jj}$$

and so, if a diagonal element a_{ii} vanishes in a positive matrix, every element in the same row and column is 0.

Let the positive matrix A correspond to a linear transformation in the Hilbert space which we denote by the same letter, and let v_k be the system of eigenvectors for A. We will have $Av_k=\lambda_k v_k$ with $\lambda_k \geqq 0$. Define a new transformation B by setting μ_k equal to the positive square root of λ_k and putting $Bv_k=\mu_k v_k$. Evidently B is a transformation for which $B^2=A$. Thus we will have $a_{ij}=(Ae_j, e_i)=(Be_j, Be_i)$ making use of the fact that B is symmetric. We finally set $g_k=Bv_k$ to obtain

$$a_{ij}=(g_j, g_i).$$

It follows that A is a Gram's matrix. On the other hand, any Gram's matrix is a positive matrix:

$$\sum\sum (g_j, g_i) z_i \bar{z}_j = \|\sum \bar{z}_j g_j\|^2 \geqq 0.$$

Suppose that our positive matrix A is of order l and that it is written as a Gram's matrix $a_{ij}=(g_j, g_i)$. It is then easy to show that k is the dimension of the null space of A, i.e. that 0 is an eigenvalue of A with multiplicity k if and only if $l-k$ is the rank of the matrix, or, equivalently, the system of vectors g_i spans an $l-k$ dimensional subspace of the corresponding Hilbert space.

It is obvious that the positive matrices form a convex cone; if A and B are positive matrices of order l and a and b positive numbers, then $aA+bB$ is also a positive matrix of order l. It is perhaps more important to notice the following: suppose z_i is a system of numbers such that

$$\sum\sum(aA+bB)_{ij}z_i\bar{z}_j=0,$$

then, simultaneously,

$$\sum\sum a_{ij}z_i\bar{z}_j=\sum\sum b_{ij}z_i\bar{z}_j=0.$$

If follows that if we add to a positive matrix any other positive matrix with non-zero eigenvalues, the sum does not have 0 as an eigenvalue.

Let $\mathscr{H}_l$ be the l-dimensional Hilbert space and f a normalized vector in it. We will have $f=\sum a_k e_k$ relative to some orthonormal basis e_k and $1=\|f\|^2=\sum|a_k|^2$. We consider the one-dimensional projection associated with f, namely, the operator P defined as follows:

$$Pu=(u,f)f.$$

It is easy to see that P is a positive operator: $(Pu,u)=|(u,f)|^2\geqq 0$. We compute the matrix element:

$$P_{ij}=(Pe_j,e_i)=(e_j,f)(f,e_i)=\bar{\alpha}_j\alpha_i, \qquad \alpha_i=(f,e_i).$$

The matrix thus has a particularly simple form relative to the (arbitrary) basis e_k. Conversely, any matrix of the form $a_{ij}=\beta_i\bar{\beta}_j$ is of the form cP where P is a one-dimensional projection, and c, the trace of the matrix, is a suitable positive constant.

We pass now to the class of symmetric matrices of order l; these form a vector space over the real scalars with dimension l^2. On that space we define the Schur product as follows. If A and B are two symmetric matrices, their Schur product $A\circ B$ is the matrix C where

$$C_{ij}=A_{ij}B_{ij}.$$

Thus the elements of the matrices are multiplied pointwise. It is obvious that the Schur product is distributive and commutative:

$$A\circ(B+C)=A\circ B+A\circ C, \qquad A\circ B=B\circ A$$

and that

$$A\circ(\lambda B)=(\lambda A)\circ B=\lambda(A\circ B)$$

for any scalar λ. It should be emphasized that the Schur product is defined for matrices and not for transformations.

The important theorem is due to I. Schur.

Theorem. *The Schur product of two positive matrices is again a positive matrix.*

Proof. Let the matrices in question be A and B. Consider the transformation A: from the spectral theorem we can write A in the form

$$A=\sum_{k=0}^{l} \lambda_k P_k$$

where each λ_k is a positive eigenvalue of A and each P_k is the one-dimensional projection on the corresponding normalized eigenvector v_k. Similarly

$$B=\sum_{j=0}^{l} \mu_j Q_j$$

where each $\mu_j \geqq 0$ is an eigenvalue of B, and Q_j the corresponding one-dimensional projection. We consider all transformations as written relative to the same basis $\{e_k\}$ in such a way that the transformations A and B are represented by the matrices A and B whose Schur product we are to compute. Now, owing to the distributivity, the matrix $A \circ B$ is given by the equation

$$A \circ B = \sum\sum \lambda_k \mu_j P_k \circ Q_j$$

in which every coefficient is non-negative. Since the set of positive matrices form a convex cone, it is only necessary to show that the Schur products $P_k \circ Q_j$ are positive matrices. For this, we may omit the subscripts k and j and note that P and Q have the form

$$P_{ij}=\alpha_i \bar{\alpha}_j, \qquad Q_{ij}=\beta_i \bar{\beta}_j$$

so $P \circ Q$ has the form $\alpha_i \beta_i \overline{\alpha_j \beta_j}$. Accordingly, $P \circ Q$ is a positive matrix, as desired.

Corollary. *If a_{ij} is a positive matrix, the matrix $k_{ij}=e^{a_{ij}}$ is also a positive matrix.*

Proof. Evidently we may write K as a convergent series

$$K=I+A+\frac{A \circ A}{2!}+\frac{A \circ A \circ A}{3!}+\cdots$$

every term of which is a positive matrix by Schur's Theorem. It follows that the partial sums of the series are positive matrices and therefore that K itself is the limit of a sequence of positive matrices. Accordingly K is a positive matrix.

The operator K introduced in the Corollary is not the exponential of A as an operator function, although that exponential is also a positive matrix. We should rather call K the Schur exponential of A.

In the sequel we will have occasion to make use of the matrix E, every element of which equals $+1$. Clearly, E is a positive matrix of the form cP where c is the trace of the matrix and P a projection onto the normalized vector, all of whose components are equal.

The following theorem gives a criterion for positivity which we will invoke frequently in the sequel.

Theorem. *Let A be a symmetric matrix of order n such that the n minors*

$$a_{11}, \quad \det\begin{bmatrix} a_{11} & a_{12} \\ a_{21} & a_{22} \end{bmatrix}, \quad \det\begin{bmatrix} a_{11} & a_{12} & a_{13} \\ a_{21} & a_{22} & a_{23} \\ a_{31} & a_{32} & a_{33} \end{bmatrix}, \ldots, \det A$$

are all strictly positive; then A is a positive matrix.

Proof. We argue by induction on n. The case $n=1$ is trivial, while for $n=2$ we note that $a_{11}>0$ by hypothesis, and $a_{11}a_{22}-|a_{12}|^2>0$ makes $a_{22}>0$. Accordingly, both the trace and the determinant of A are positive, that is, the sum and product of the eigenvalues is positive. Evidently both eigenvalues are positive and A is a positive matrix.

For the passage from n to $n+1$ we write the eigenvalues of A in increasing order: $\lambda_1 \leqq \lambda_2 \leqq \lambda_3 \leqq \cdots \leqq \lambda_{n+1}$ and let e_i denote the corresponding normalized eigenvectors. Since

$$\det A = \lambda_1 \lambda_2 \lambda_3 \ldots \lambda_{n+1}$$

is positive by hypothesis, no eigenvalue is 0 and the number of negative eigenvalues is even. Hence, if A is not a positive matrix, both λ_1 and λ_2 are negative. The quadratic form (Au, u) is therefore negative on the two dimensional subspace spanned by e_1 and e_2. However, there is always a normalized vector of the form $u=ae_1+be_2$ having the component $u_{n+1}=0$. For this vector

$$(Au, u) = \sum_{i=1}^{n} \sum_{j=1}^{n} a_{ij} u_i \bar{u}_j$$

is positive by the inductive hypothesis. From this contradiction it follows that A is a positive matrix.

4. Regularizations

Let $\varphi(x)$ be a C^∞-function defined on the real axis, vanishing outside the closed interval $|x| \leqq 1$. We will also require that $\varphi(x)$ is positive and even: $\varphi(x)=\varphi(-x) \geqq 0$ and that the function is so normalized that

$$\int_{-1}^{1} \varphi(x)\,dx = 1\,.$$

Now, given any locally integrable function $f(x)$ on the open interval (a, b) we form its regularization of order ε where the positive ε is small:

$$f_\varepsilon(x) = \frac{1}{\varepsilon} \int \varphi\left(\frac{x-y}{\varepsilon}\right) f(y)\,dy .$$

This may also be written $\int \varphi(z) f(x-\varepsilon z)\,dz$. It is easy to see that the value of the regularization $f_\varepsilon(x)$ is an average of the values taken by the function f in an ε-neighborhood of the point x. Accordingly, if x_0 is a point of continuity of the function, the regularizations converge with decreasing ε to $f(x_0)$. If the function is continuous it is then uniformly continuous on any closed subinterval, and the regularizations converge uniformly on such subintervals.

Since the function $\varphi(x)$ above is C^∞, the regularizations $f_\varepsilon(x)$ are themselves C^∞, since we may differentiate with respect to x under the integral sign. It is exactly this smoothness of the regularizations which makes them so useful.

In almost every case where we shall make use of the regularizations we will be regularizing a function $f(x)$, continuous on the open interval (a, b). The regularizations will then make sense on intervals of the form $(a+\varepsilon, b-\varepsilon)$.

There is another very important property of the regularizations: the regularization of a derivative is the derivative of the regularization. Suppose, for example, that the function $f(x)$ above has a continuous derivative $f'(x)$. If we write its regularization we obtain an integral which we integrate by parts:

$$\begin{aligned}(f')_\varepsilon(x) &= \frac{1}{\varepsilon} \int \varphi\left(\frac{x-y}{\varepsilon}\right) f'(y)\,dy = \frac{-1}{\varepsilon} \int \frac{d}{dy} \varphi\left(\frac{x-y}{\varepsilon}\right) f(y)\,dy \\ &= \frac{1}{\varepsilon} \int \frac{d}{dx} \varphi\left(\frac{x-y}{\varepsilon}\right) f(y)\,dy = \frac{d}{dx} f_\varepsilon(x) .\end{aligned}$$

It should be noted that the full strength of the hypothesis that $f'(x)$ was continuous was not used; all that was necessary was the integration by parts and this would have been legitimate if we merely required that $f(x)$ be absolutely continuous on closed subintervals of (a,b). Our next theorem, however, does not even require that.

Theorem. *Let $f(x)$ be monotone in (a, b); then $(d/dx) f_\varepsilon(x)$ converges to $f'(x)$ with decreasing ε for every x where that derivative exists and is finite.*

Proof. We may suppose that 0 is a point of (a, b) where the derivative exists and is finite, and therefore, in a neighborhood of that point

$$f(x) = M + mx + \eta(x)$$

where $M = f(0)$, $m = f'(0)$ and $\eta(x)$ is $o(|x|)$. We form the regularization of order ε and differentiate at $x=0$ to obtain

$$\frac{d}{dx} f_\varepsilon(0) = \frac{1}{\varepsilon^2} \int \varphi'\left(\frac{-y}{\varepsilon}\right) [M + m y + \eta(y)]\, dy.$$

This integral is the sum of three integrals, the first of which vanishes since $\varphi'(y)$ is an odd function. The second integral is easily computed:

$$m \frac{1}{\varepsilon^2} \int \varphi'\left(\frac{-y}{\varepsilon}\right) y\, dy = -m \int_{-1}^{+1} \varphi'(t)\, t\, dt = m \int_{-1}^{+1} \varphi(t)\, dt = m.$$

Finally, the third integral is bounded by

$$\int_{-1}^{+1} |\varphi'(t)|\, dt \frac{1}{\varepsilon} \sup_{|y|<\varepsilon} |\eta(y)|$$

and this bound converges to 0 with diminishing ε in view of the behaviour of $\eta(x)$.

5. Completely Monotone Functions

Let $f(x)$ be real and C^∞ on the half-axis $x>0$; we shall say that the function is completely monotone if for all $n \geqq 0$

$$(-1)^n f^{(n)}(x) \geqq 0$$

on the half-axis. The completely monotone functions are characterized by a theorem of S. N. Bernstein, which we prefer to divide into two parts.

Little Bernstein Theorem. *If $f(x)$ is completely monotone, then it is the restriction to the half-axis of a function analytic in the right half-plane.*

Proof. Choose $b>0$ and consider the formal Taylor expansion of $f(x)$ about the point b:

$$\sum_{n=0}^{\infty} \frac{f^{(n)}(b)}{n!} (x-b)^n.$$

We note that every term in this series is positive for x in the interval $(0, b)$ and so the partial sums $S_N(x)$ form a monotone increasing sequence over that interval. Those partial sums are uniformly bounded; indeed, from the Remainder formula for the Taylor expansion we have

$$f(x) - S_N(x) = \frac{f^{(N+1)}(\xi)}{(N+1)!} (x-b)^{(N+1)}$$

where ξ is appropriately chosen in $(0,b)$ and this is a positive quantity. Hence the partial sums converge, and the formal Taylor series converges in the interval $(0, b)$. To show that it converges to the right value, viz. $f(x)$, we note first that the functions $|f^{(N)}(x)|$ are decreasing, hence, if we consider x in the interval $(3b/4, b)$ the remainder can be estimated as follows:

$$f(x)-S_N(x)=\frac{f^{(N+1)}(\xi)}{(N+1)!}(x-b)^{N+1}\leqq\frac{|f^{(N+1)}(b/2)|}{(N+1)!}(b/4)^{N+1}.$$

The quantity in the right converges to 0 with increasing N since it is the general term of a convergent series, viz. the expansion of $f(x)$ about $b/2$ taken at the point $x=b/4$. It follows that the series converges to $f(x)$ for $3b/4<x<b$ and since b is arbitrary, $f(x)$ is analytic in a neighborhood of the positive real axis. Accordingly the function is analytic in any circle with center $b>0$ and of radius b. The union of all such circles is the open right half-plane. This completes the proof.

Big Bernstein Theorem. *If $f(x)$ is completely monotone, then it is the restriction to the half-axis of the Laplace transform of a positive measure.*

Proof. We first suppose that $f(0)=1$ and note that for any sequence of positive numbers $\{\lambda_k\}$ we have

$$(-1)^{n-1}[\lambda_1,\lambda_2,\ldots,\lambda_n]_f\geqq 0.$$

We select an unbounded increasing sequence of positive numbers $\{\lambda_n\}$ so chosen that the series $\sum 1/\lambda_n$ is divergent, and construct the corresponding system of Newtonian interpolation polynomials $P_n(x)$. It must first be shown that $f(x)$ can be represented by a series of the form

$$\sum_{k=0}^{\infty}(-1)^k[\lambda_1,\lambda_2,\ldots,\lambda_{k+1}](-1)^k P_k(x)$$

where, as in the future, we omit the subscript f on the divided differences. First consider the series on the interval $(0,\lambda_1)$. For x in this interval, the functions $(-1)^k P_k(x)$ are all positive, and we have already noticed that the coefficients of these polynomials are positive. Accordingly, the series is a series of positive terms and the partial sums form a monotone increasing sequence of functions on the interval $(0,\lambda_1)$. These partial sums are uniformly bounded, since we may write

$$\begin{aligned} f(x)-\sum_{k=0}^{n-1}(-1)^k[\lambda_1,\lambda_2,\ldots,\lambda_{k+1}](-1)^k P_k(x)&=R_{n-1}(x)\\ &=(-1)^n[\lambda_1,\lambda_2,\ldots,\lambda_n,x](-1)^n P_n(x)\geqq 0. \end{aligned}$$

It follows that the series converges on the interval $(0, \lambda_1)$. Virtually the same argument shows that the series converges on the interval (λ_1, λ_2). Here, for all $k>1$, $(-1)^k P_k(x)$ is negative, although the coefficients of these polynomials are positive. Thus the partial sums, after perhaps the first term, form a monotone decreasing sequence. The remainder term is negative, so the diminishing sequence is uniformly bounded from below by $f(x)$. The series therefore converges in (λ_1, λ_2). In general, in the interval $(\lambda_j, \lambda_{j+1})$ the polynomials $(-1)^k P_k(x)$ have the same algebraic sign as $(-1)^j$ provided $k>j$, and so the partial sums of order $>j$ form a monotone sequence. This sequence converges since the remainder term has the same sign as the terms of the series. Accordingly the series converges for every real x. It remains to show that it converges to the right sum, i.e., $f(x)$.

Since the series converges for $x=0$ we have

$$1=f(0)=[\lambda_1]+\sum_{k=1}^{\infty}(-1)^k[\lambda_1, \lambda_2, \ldots, \lambda_{k+1}]\prod_{j=1}^{k}\lambda_j .$$

We have also

$$R_{n-1}(x)=(-1)^n[\lambda_1, \lambda_2, \ldots, \lambda_n, x]\prod_{j=1}^{n}(\lambda_j-x)$$

and so infer that there exists C_M so that for $x<M$

$$|R_{n-1}(x)|\leqq(-1)^n[\lambda_1, \lambda_2, \ldots, \lambda_n, x]\prod_{j=1}^{n}\lambda_j C_M .$$

Now we multiply by x/λ_n to obtain

$$\left|\frac{x}{\lambda_n}R_{n-1}(x)\right|\leqq(-1)^n[\lambda_1, \lambda_2, \ldots, \lambda_n, x]\,x\prod_{j=1}^{n-1}\lambda_j C_M .$$

Adjoin x to the sequence $\{\lambda_n\}$ and suppose that x occurs in the interval $(\lambda_N, \lambda_{N+1})$. With this new sequence we form the appropriate series representation of $f(x)$ and compute the value of that function at $x=0$. The terms of order greater than N in the resulting series will have the form

$$(-1)^n[\lambda_1, \lambda_2, \ldots, \lambda_n, x]\,x\prod_{j=1}^{n-1}\lambda_j$$

and from the convergence of this series we infer the convergence of the sum

$$\sum_{n=1}^{\infty}|x R_{n-1}(x)/\lambda_n| .$$

Because $\sum 1/\lambda_n$ diverges, it follows that $R_{n-1}(x)$ converges to 0 with increasing n and therefore that the interpolation series converges to the sum $f(x)$.

Now it is important to notice that the series representation of $f(x)$ which we have obtained involved only the fact that the $\{\lambda_k\}$ formed an unbounded monotone sequence such that the series of reciprocals diverged. Thus, were we to neglect the first $N-1$ points of that sequence, the function $f(x)$ would admit a convergent expansion in terms of the new sequence and the corresponding interpolation polynomials. We write this expansion explicitly as follows:

$$f(x)=[\lambda_N]+\sum_{k=1}^{\infty}(-1)^k[\lambda_N,\lambda_{N+1},\ldots,\lambda_{N+k}]\prod_{j=0}^{k-1}\lambda_{N+j}\prod_{j=0}^{k-1}(1-x/\lambda_{N+j}).$$

Now for each integer $N\geqq 1$ we introduce a positive measure μ_N defined on the real axis as follows. The measure μ_N puts the mass $[\lambda_N]$ at the origin, and at the k-th point of the sequence

$$\tau_{N,k}=\frac{1}{\lambda_N}+\frac{1}{\lambda_{N+1}}+\frac{1}{\lambda_{N+2}}+\cdots+\frac{1}{\lambda_{N+k-1}}$$

puts the mass

$$\mu_{N,k}=(-1)^k[\lambda_N,\lambda_{N+1},\ldots,\lambda_{N+k}]\prod_{j=0}^{k-1}\lambda_{N+j}.$$

The measure μ_N is a positive measure of finite total mass and its Laplace transform is easily found.

$$\hat{\mu}_N(x)=[\lambda_N]+\sum_{k=1}^{\infty}\mu_{N,k}e^{-\tau_{N,k}x}$$

while

$$f(x)=[\lambda_N]+\sum_{k=1}^{\infty}\mu_{N,k}\prod_{j=0}^{k-1}(1-x/\lambda_{N+j}).$$

Our object is now to show that the Laplace transforms $\hat{\mu}_N(x)$ converge with increasing N to $f(x)$. We consider the behavior of these functions on some fixed interval $(0,M)$. For a given small ε choose N so large that $M/\lambda_N<\varepsilon/2$: it will then follow that for all x in the interval $(0,M)$ and all $j\geqq 0$

$$-(1+\varepsilon)\frac{x}{\lambda_{N+j}}\leqq\log\left(1-\frac{x}{\lambda_{N+j}}\right)\leqq-\frac{x}{\lambda_{N+j}}.$$

Summing from $j=0$ to $j=k-1$ we find

$$-\tau_{N,k}(1+\varepsilon)x\leqq\log\prod_{j=0}^{k-1}\left(1-\frac{x}{\lambda_{N+j}}\right)\leqq-\tau_{N,k}x$$

and therefore

$$e^{-\tau_{N,k}(1+\varepsilon)x}\leqq\prod_{j=0}^{k-1}\left(1-\frac{x}{\lambda_{N+j}}\right)\leqq e^{-\tau_{N,k}x}.$$

We finally infer that for such large values of N, with x in the interval $(0, M)$,

$$\hat{\mu}_N((1+\varepsilon)x) \leqq f(x) \leqq \hat{\mu}_N(x).$$

We may let N increase and ε diminish in this inequality to obtain

$$\limsup_N \hat{\mu}_N(x) \leqq f(x) \leqq \liminf_N \hat{\mu}_N(x)$$

and therefore to infer that on any interval of the form $(0, M)$ $f(x)$ is the limit of a sequence of Laplace transforms of positive measures.

We next map the half-axis $t>0$ onto the interval $(0, 1)$ by the substitution $y=e^{-t}$, carrying the measures μ_N on the half-axis over into positive measures ν_N on the closed interval. We may write the Laplace transform in the form

$$\hat{\mu}_N(x) = \int_0^1 y^x \, d\nu_N(y)$$

and note that the measures ν_N all have total mass $1=f(0)$. If

$$p(y) = \sum a_i y^i$$

is any polynomial in y, then the numbers

$$\int_0^1 p(y) \, d\nu_N(y) = \sum a_i \hat{\mu}_N(i)$$

converge with increasing N to $\sum a_i f(i)$, and it follows from Helly's theorem that the measures ν_N converge weakly to a positive measure for which

$$f(x) = \int_0^1 y^x \, d\nu(y).$$

From the continuity of $f(x)$ at $x=0$ it can be shown that the measure ν puts no mass at the point $y=0$. We then transfer the measure back to the right half-axis by the mapping $t=-\log y$ to obtain a positive measure μ with Laplace transform $f(x)$.

It remains to extend the argument to the case where the special hypothesis $f(0)=1$ need no longer hold. In this case we form the ratio $f(x+\varepsilon)/f(\varepsilon)$ for some small positive ε to obtain a completely monotone function which is a Laplace transform. We write

$$f(x+\varepsilon) = f(\varepsilon) \int e^{-xt} \, d\mu(t)$$

for some positive measure μ of finite total mass. We infer immediately that

$$f(x) = \int e^{-xt} e^{\varepsilon t} f(\varepsilon) \, d\mu(t)$$

and obtain $f(x)$ as the Laplace transform of the positive measure $e^{\varepsilon t} f(\varepsilon) \, d\mu(t)$. This completes the proof of the Big Bernstein Theorem.

Chapter II. Pick Functions

1. Definitions and Examples

By the Pick class we shall mean the class of functions $\varphi(\zeta)=U(\zeta)+iV(\zeta)$ analytic in the upper half-plane with positive imaginary part; thus if $\zeta=\xi+i\eta$ then $V(\zeta)\geqq 0$ if $\eta>0$. We denote the class by the letter P, and sometimes call functions in the class Pick functions.

The Pick functions evidently form a convex cone; i.e. if a and b are positive numbers and $\varphi_1(\zeta)$ and $\varphi_2(\zeta)$ two functions in P, the function $a\varphi_1(\zeta)+b\varphi_2(\zeta)$ is also a Pick function. The class is also closed under composition: the composed function $(\varphi_2\circ\varphi_1)(\zeta)=\varphi_2(\varphi_1(\zeta))$ is analytic for $\eta>0$ and has positive imaginary part there. Since the function $-1/\zeta$ is in P, it follows that whenever $\varphi(\zeta)$ is in P, so also is $-1/\varphi(\zeta)$.

Let $\varphi(\zeta)=U(\zeta)+iV(\zeta)$ be a Pick function which is real at some ζ in the upper half-plane. It is clear that the positive and harmonic function V then vanishes at that point and therefore, from the mean value property of harmonic functions, that $V(\zeta)=0$ identically. Thus $\varphi(\zeta)$ is a real constant. We see that a non-trivial Pick function is never real in the open upper half-plane.

Let (a, b) be an open interval of the real axis; by $P(a, b)$ we denote the subclass of P consisting of those Pick functions which admit an analytic continuation across the interval into the lower half-plane and where the continuation is by reflection. Thus the functions in this class are real on the interval (a, b) and are continuable throughout the lower half-plane. It is clear that $P(a,b)$ is also a convex cone. A non-constant function in that class takes real values only on the real axis. If $\varphi(\zeta)=U(\zeta)+iV(\zeta)$ is in $P(a, b)$ then $V(x)=0$ for $a<x<b$, hence, $V(x+iy)-V(x)>0$ for $y>0$ and therefore $V_y(x)\geqq 0$. From one of the Cauchy-Riemann equations we find $U_x(x)=V_y(x)\geqq 0$ and therefore $\varphi(x)=U(x)$ is an increasing function on the interval.

It is worthwhile to carry this type of argument a little further. Suppose that $\varphi(\zeta)$ is a non-constant rational function in the class $P(a, b)$. It is then evident that the function takes real values on the real axis and only there. It is also not hard to see that there exists a real constant λ

such that $\varphi(\zeta)-\lambda$ has N distinct and simple zeros, since we have only to choose λ outside the finite set of values assumed by φ where the rational derivative vanishes. N, of course, is the degree of the rational $\varphi(\zeta)$. Between any two zeros of the Pick function $\varphi-\lambda$ there must exist a pole since the function is always increasing. This is also the case in a neighborhood of infinity. Thus there are N distinct poles, necessarily simple. It is also clear that the residue at each pole is negative. Of course one of the poles may be at infinity, in which case the corresponding term in the rational function looks like $\alpha\zeta$ with $\alpha>0$.

It is advisable to consider a few examples of Pick functions before going farther. Our arguments about rational functions in classes $P(a,b)$ make it clear that any rational Pick function, real on an interval of the axis, is necessarily of the following form:

$$\varphi(\zeta)=\alpha\zeta+\beta+\sum_{i=1}^{n}\frac{m_i}{\lambda-\zeta}$$

where $\alpha>0$, $m_i>0$ and β is real. We shall later see that this formula may be generalized to one representing all Pick functions.

Consider that determination of the square root which is positive on the right half-axis: the function so obtained is a Pick function since the argument of $\varphi(\zeta)=\sqrt{\zeta}$ is one half the argument of ζ and so the number $\varphi(\zeta)$ is in the upper half-plane, indeed, the number $\varphi(\zeta)$ falls in the upper right quadrant of the complex plane.

These considerations also make it clear that for $0<\gamma<1$ the function $\varphi(\zeta)=\zeta^{\gamma}$ is a Pick function.

We obtain another Pick function when we consider that determination of the logarithm which is real on the right half-axis: in the upper half-plane the imaginary part of the logarithm takes values in the interval $(0,\pi)$.

Another fundamental function in the Pick class is the tangent. To show that this is a Pick function we make use of the familiar addition theorem for the tangent

$$\tan(\xi+i\eta)=\frac{\tan\xi+\tan i\eta}{1-\tan\xi\tan i\eta}$$

and the identity $\tan(i\eta)=i\tanh\eta$ to obtain

$$\tan\zeta=\frac{\tan\xi+i\tanh\eta}{1-i\tan\xi\tanh\eta}.$$

It follows that the imaginary part is given by $\tanh\eta(1+\tan^2\xi)/(1+\tan^2\xi\tanh^2\eta)$ and this is positive if η is.

A more sophisticated, but very important example of a Pick function, arises in the study of self-adjoint transformations in Hilbert space.

Let H be a self-adjoint transformation; its resolvent $R_\zeta=(H-\zeta I)^{-1}$ is always a well-defined bounded operator when ζ is not on the real axis. It is well known that R_ζ depends analytically on ζ, hence, for any u in the Hilbert space, the function $\varphi(\zeta)=(R_\zeta u, u)$ is analytic in the upper half-plane. We assert that it is a Pick function. If $R_\zeta u=v$, then $(H-\zeta I)v=u$ and therefore

$$\varphi(\zeta)=(v, Hv-\zeta v)=(v, Hv)-\bar{\zeta}(v, v).$$

Since (v, Hv) is real, the imaginary part is $\eta\|v\|^2$ and this has the same sign as η.

2. The Integral Representation

The functions in the Pick class admit a canonical integral representation to which we now turn.

Theorem I. *A function $\varphi(\zeta)$ in the class P has a unique canonical representation of the form*

$$\varphi(\zeta)=\alpha\zeta+\beta+\int\left[\frac{1}{\lambda-\zeta}-\frac{\lambda}{\lambda^2+1}\right]d\mu(\lambda) \tag{1}$$

where $\alpha\geqq 0$, β is real, and $d\mu(\lambda)$ a positive Borel measure on the real λ-axis for which $\int(\lambda^2+1)^{-1}d\mu(\lambda)$ is finite. Conversely, any function of this form is in P.

There is a corresponding result for functions positive and harmonic in the half-plane.

Theorem II. *Any function $V(\xi,\eta)$ positive and harmonic in the half-plane $\eta>0$ admits a unique canonical representation of the form*

$$V(\xi,\eta)=\alpha\eta+\int\frac{\eta\, d\mu(\lambda)}{(\lambda-\xi)^2+\eta^2} \tag{2}$$

where $\alpha\geqq 0$ and $d\mu(\lambda)$ is a positive Borel measure on the real axis for which $\int(\lambda^2+1)^{-1}d\mu(\lambda)$ is finite. Conversely, any function of this form is positive and harmonic in the half-plane.

Before going through the proof, let us remark that either of these theorems can be deduced from the other. To prove Theorem II, for example, we may suppose $V(\xi,\eta)$ given, harmonic and positive in the half-plane. Such a function has a harmonic conjugate $U(\xi,\eta)$ which is determined up to an additive real constant and which is harmonic throughout the half-plane. Accordingly, the analytic function $\varphi(\zeta)=U(\zeta)+iV(\zeta)$ is in P and admits the representation (1). The function $\varphi(\zeta)$

has been determined up to an additive real constant, and so only the value of β in the formula is undetermined by $V(\zeta)$. If we take the imaginary part of φ in the representation (1) we obtain $V(\zeta)$ in the representation (2). This representation is unique since the value of β does not occur in the formula. On the other hand, Theorem I is obtained from Theorem II by writing $V(\zeta)$, the imaginary part of $\varphi(\zeta)$ in the form (2) and noting that the function defined by (1) and the given values of α and μ with $\beta=\operatorname{Re}[\varphi(i)]$ is an analytic function in the half-plane with the given imaginary part and the correct value at $\zeta=i$. Hence (1) is a representation of φ in P. The representation is unique since $V(\zeta)$, by Theorem II, completely determined the values α and μ.

Two completely analogous theorems hold for functions in the unit disk.

Theorem III. *A function $f(z)=u(z)+iv(z)$ analytic in the disk $|z|<1$ with positive real part there admits a unique canonical representation of the form*

$$f(z)=\int_0^{2\pi}\frac{e^{i\theta}+z}{e^{i\theta}-z}\,d\omega(\theta)+iv(0) \tag{3}$$

where $v(0)$ is real and ω a positive Borel measure on the interval $[0,2\pi]$ with finite total mass. Conversely, any function of the form (3) is analytic in the disk $|z|<1$ and has a positive real part there.

Theorem IV. *A function $u(z)$ harmonic and positive in $|z|<1$ has a unique canonical representation of the form*

$$u(z)=u(re^{i\theta})=\int_0^{2\pi}\frac{1-r^2}{1+r^2-2r\cos(\theta-\phi)}\,d\omega(\phi) \tag{4}$$

where ω is a positive Borel measure on the interval $[0,2\pi]$ with finite total mass. Conversely, any function of the form (4) is harmonic and positive in the circle.

The equivalence of Theorems III and IV is established again by using the fact that the harmonic conjugate is determined up to an additive real constant. Here we must note also that for $z=re^{i\phi}$

$$\frac{e^{i\theta}+z}{e^{i\theta}-z}=\frac{1-r^2-i2r\sin(\theta-\phi)}{1+r^2-2r\cos(\theta-\phi)}.$$

We shall next show the equivalence of Theorems I and III, making use of the functions

$$\zeta(z)=\frac{1}{i}\,\frac{z+1}{z-1}\quad\text{and}\quad z(\zeta)=\frac{\zeta-i}{\zeta+i},$$

a pair of linear fractional transformations which are inverses of one another. The function $\zeta(z)$ maps the unit disk onto the upper half-plane, while $z(\zeta)$ carries the half-plane back onto the unit disk. If $f(z)=u(z)+iv(z)$ is analytic in the disk $|z|<1$ with positive real part, the function $\varphi(\zeta) = if(z(\zeta))$ is in the class P; conversely, if $\varphi(\zeta)$ is in P, the function $f(z) = -i\varphi(\zeta(z))$ is analytic in the disk with positive real part there. Thus the two classes of functions are in a one-to-one correspondence. We can therefore compute the canonical representation of $\varphi(\zeta)$ in P from that of the corresponding $f(z)$ as follows. We suppose $f(z)$ given by (3). If the measure $d\omega$ has a point mass $\alpha>0$ at $\theta=0$, we separate out that mass, writing (3) more particularly:

$$f(z)=\frac{1+z}{1-z}\alpha-i\beta+\int\frac{e^{i\theta}+z}{e^{i\theta}-z}\,d\omega'(\theta)$$

where $d\omega'$ is the measure obtained from $d\omega$ by omitting the point mass α at $\theta=0$. We now form $if(z(\zeta))$; the first two terms reduce to $\alpha\zeta+\beta$. In changing variables in the integral, we have only to compose the transformation $z(\zeta)$ with $(e^{i\theta}+z)/(e^{i\theta}-z)$ and multiply by i. We obtain for the integral, then

$$\int\frac{\zeta\cos\theta/2-\sin\theta/2}{\zeta\sin\theta/2+\cos\theta/2}\,d\omega'(\theta)$$

and introduce the change of variables $\lambda=-\cot\theta/2$ which carries the circle onto the real axis, mapping the deleted point $z=1$ into infinity. The measure $d\omega'$ is thus brought over to a measure of finite total mass on the real axis, and we have

$$\varphi(\zeta)=\alpha\zeta+\beta+\int\frac{\lambda\zeta+1}{\lambda-\zeta}\,d\nu(\lambda) \quad \text{where} \quad \int d\nu(\lambda)<+\infty\,. \tag{5}$$

Writing, finally, $d\mu(\lambda)=(\lambda^2+1)\,d\nu(\lambda)$, we obtain (1).

It should be emphasized that the whole argument is reversible: if a function $\varphi(\zeta)$ in P has the representation (1), $f(z)=-i\varphi(\zeta(z))$ then has the form (3). Thus the theorems which we have stated are all equivalent, and it will be enough to prove IV. For this, we suppose at first that the function $u(z)=u(re^{i\theta})$ is positive and harmonic in a disk of radius greater than 1; it is then bounded and continuous on $|z|=1$. We may determine the harmonic conjugate $v(z)$ in such a way that it vanishes at $z=0$; now $f(z)=u(z)+iv(z)$ is analytic in a disk of radius greater than 1 and is there represented by the power series $\sum a_n z^n$ which converges absolutely and uniformly on the circle $|z|=1$. The real part $u(z)$ is then given by a series:

$$u(z)=\frac{f(z)+\overline{f(z)}}{2}=\frac{1}{2}\sum_{n=1}^{\infty}\left[a_n z^n+\bar{a}_n \bar{z}^n\right]+a_0$$

and this series also converges absolutely and uniformly for $|z| \leqq 1$. Thus, our function admits a representation

$$u(r,\theta)=\sum_{-\infty}^{+\infty} A_n r^{|n|} e^{in\theta}$$

and it is easy to see that the A_n are the Fourier coefficients of the function $u(1,\theta)$. Accordingly

$$u(r,\theta)=\frac{1}{2\pi}\sum_{-\infty}^{+\infty} r^{|n|} e^{in\theta}\int_0^{2\pi} u(e^{i\phi})e^{-in\phi}\,d\phi$$

and for $r<1$ we can interchange the order of integration and summation to obtain

$$\frac{1}{2\pi}\int_0^{2\pi}\sum r^{|n|} e^{in(\theta-\phi)}u(e^{i\phi})\,d\phi$$

which reduces to

$$u(z)=u(r,\theta)=\frac{1}{2\pi}\int_0^{2\pi}\frac{1-r^2}{1+r^2-2r\cos(\theta-\phi)}u(e^{i\phi})\,d\phi\,.$$

This is the representation (4) for $u(z)$ where the measure $d\omega(\phi)$ is given by the positive and bounded density $(1/2\pi)u(e^{i\phi})$ relative to Lebesgue measure on the interval $[0,2\pi]$. It is important to note that the total mass of this measure is just the value of $u(z)$ at the origin:

$$u(0)=\frac{1}{2\pi}\int_0^{2\pi} u(e^{i\phi})\,d\phi\,.$$

The general case is an immediate consequence of this one: if $u(z)$ is positive and harmonic in $|z|<1$, the function $u_\varepsilon(z)=u(z/(1+\varepsilon))$ is positive and harmonic in the disk $|z|<1+\varepsilon$ and there admits a representation (4) with a positive measure $d\omega_\varepsilon(\theta)$ of total mass $u_\varepsilon(0)=u(0)$. As the positive ε approaches 0, the functions $u_\varepsilon(z)$ converge to $u(z)$ uniformly on compact subsets of $|z|<1$, while the system of positive measures $d\omega_\varepsilon$ satisfies the hypothesis of Helly's theorem, since there is a uniform bound for their masses. Thus there exists a sequence ε_n converging to 0 corresponding to a sequence of measures $d\omega_n$ converging weakly to a positive measure $d\omega$ such that

$$u(z)=\lim u_\varepsilon(z)=\int \frac{1-r^2}{1+r^2-2r\cos(\theta-\phi)}\, d\omega(\phi)\,.$$

We have yet to establish the uniqueness of the canonical representations given by these theorems. It will be enough to do this for Theorem I. We have to show that the elements α, β and $d\mu$ of the canonical representation are uniquely determined by the function $\varphi(\zeta)$ in P. It is obvious that $\beta=\mathrm{Re}[\varphi(i)]$ and therefore that φ determines β. To show that α is determined, we use the representation (5) which has the advantage of displaying the measure $d\nu(\lambda)$ of finite total mass and compute the limit as the positive η approaches infinity of the ratio $\varphi(i\eta)/i\eta$. The ratio in question is equal to

$$\alpha+\frac{\beta}{i\eta}+\int \frac{\lambda^2+1+i\lambda(\eta+1/\eta)}{\lambda^2+\eta^2}\, d\nu(\lambda)\,.$$

It is clear that the integrand converges pointwise to 0 on the λ-axis as the positive η approaches infinity and it is easy to see that its real and imaginary parts are uniformly bounded by 1 on the real λ-axis when $\eta>1$. Thus the Lebesgue convergence theorem guarantees that the limit of the integral is 0. Accordingly

$$\alpha=\lim_{\eta\to\infty} \varphi(i\eta)/i\eta\,.$$

Virtually the same argument shows that α is the limit of $V(i\eta)/\eta$ as η increases without bound.

To show that the measure $d\mu$ is uniquely determined we consider the monotone increasing function $\mu(\lambda)$ defined on the whole real axis which corresponds to the measure; this function, of course, is determined up to certain normalizations. We make the conventions $\mu(0)=0$ and $\mu(\lambda)=(\mu(\lambda+0)+\mu(\lambda-0))/2$ to have the following lemma:

Lemma 1. *For any finite interval* $a<x<b$

$$\mu(b)-\mu(a)=\lim_{\eta\to 0}\frac{1}{\pi}\int_a^b V(x+i\eta)\,dx\,.$$

Proof. Suppose first that $d\mu$ has a compact support; it is therefore a measure of finite total mass. We simply compute the limit on the right:

$$\int_a^b V(x+i\eta)\,dx = \int_a^b \left[\alpha\eta + \int \frac{\eta\, d\mu(\lambda)}{(\lambda-x)^2+\eta^2}\right] dx$$

$$= \alpha\eta(b-a) + \int \int_a^b \frac{dx}{(x-\lambda)^2+\eta^2}\,\eta\, d\mu(\lambda)$$

and making the change of variable $u=(x-\lambda)/\eta$ the inner integral becomes

$$\int_{\frac{a-\lambda}{\eta}}^{\frac{b-\lambda}{\eta}} \frac{du}{u^2+1} = \operatorname{arc\,tan}\left(\frac{b-\lambda}{\eta}\right) - \operatorname{arc\,tan}\left(\frac{a-\lambda}{\eta}\right).$$

Thus for all positive η

$$\frac{1}{\pi}\int_a^b V(x+i\eta)\,dx = \frac{1}{\pi}\int \left[\operatorname{arc\,tan}\left(\frac{b-\lambda}{\eta}\right) - \operatorname{arc\,tan}\left(\frac{a-\lambda}{\eta}\right)\right]$$

$$\times d\mu(\lambda) + \alpha\eta(b-a)/\pi.$$

Evidently the last term converges to 0 with decreasing η.

For each $\eta>0$ the integrand is non-negative and bounded by π. As η decreases, the integrand converges to a function $F(\lambda)$ which vanishes outside the closed interval $[a,b]$ and equals π in the open interval (a,b). At the endpoints a and b we have $F(a)=F(b)=\pi/2$. From the Lebesgue convergence theorem, then, the limit in question is

$$\frac{1}{\pi}\int F(\lambda)\,d\mu(\lambda) = \mu(b)-\mu(a).$$

When the measure is not of finite total mass we write $\varphi(\zeta)$ in the form (1) and decompose it into the sum of two functions $\varphi_1(\zeta)+\varphi_2(\zeta) = \varphi(\zeta)$ where

$$\varphi_1(\zeta) = \alpha\zeta + \beta + \int_{-N}^{+N} \left[\frac{1}{\lambda-\zeta} - \frac{\lambda}{\lambda^2+1}\right] d\mu(\lambda)$$

and

$$\varphi_2(\zeta) = \int_{|\lambda|>N} \left[\frac{1}{\lambda-\zeta} - \frac{\lambda}{\lambda^2-1}\right] d\mu(\lambda),$$

the integer N being chosen so large that the open interval $(a-1, b+1)$ is contained in $[-N, N]$. There is then a corresponding decomposition of $V(\zeta)=V_1(\zeta)+V_2(\zeta)$ and we have only to show that $V_2(x+i\eta)$ converges to 0 uniformly on $[a, b]$. However, in that interval

$$V_2(x+i\eta)=\int\limits_{|\lambda|>N}\frac{\eta\, d\mu(\lambda)}{(\lambda-x)^2+\eta^2}\leqq\eta\left(\int\limits_{\lambda<-N}\frac{d\mu}{(\lambda-a)^2}+\int\limits_{\lambda>N}\frac{d\mu}{(\lambda-b)^2}\right)=C\eta$$

hence $V_2(x+i\eta)\leqq C\eta$ on $[a, b]$. The proof is complete.

If we pass to the special case when the function $\varphi(\zeta)$ can be continued analytically into the lower half-plane across an interval (a, b) of the real axis, then, on closed subintervals of that interval the function $V(x+i\eta)$ converges uniformly with decreasing η to a bounded and continuous function $V(x)$; the measure $d\mu(\lambda)$ then appears as $(1/\pi)V(\lambda)d\lambda$ on such a subinterval. When the analytic continuation is possible by reflection, the function $V(\lambda)$ vanishes on the subinterval and so μ has no mass on the interval. On the other hand, if μ has no mass in the interval (a, b), the integral in (1) makes sense for all $\zeta=\xi+i0$ in that interval and is clearly real there; the function can be continued through the lower half-plane by reflection, and the continued function is still given by the formula (1). We have therefore established another result.

Lemma 2. *A Pick function $\varphi(\zeta)$ belongs to $P(a, b)$ if and only if the corresponding measure μ puts no mass in the interval (a, b).*

In general a function in P has no analytic continuation across the real axis; the formula (1) makes sense in the lower half-plane and there represents an analytic function $\psi(\zeta)$ which satisfies the equation $\psi(\zeta)=\overline{\varphi(\bar{\zeta})}$ but $\psi(\zeta)$ is in no way an analytic continuation of $\varphi(\zeta)$.

The remarks which we have made earlier concerning rational functions in the Pick class, real on some interval of the real axis, could all be deduced from Theorem I. The function is necessarily in some class $P(a, b)$ and the corresponding measure is evidently concentrated at the real poles of the function. It follows that the function is of the form

$$\varphi(\zeta)=\alpha\zeta+\beta+\sum_{i=1}^{n}\frac{m_i}{\lambda_i-\zeta}\quad\text{where } m_i>0\,.$$

This also makes it clear that the poles are simple, with negative residues.

It is easy to find the canonical representation (1) for many important functions.

Example 1. The function $\varphi(\zeta)=-1/\zeta$ is rational and real on the real axis; it vanishes at infinity, and $\varphi(i)=+i$ is imaginary. Thus $\alpha=\beta=0$ and the measure $d\mu$ consists of a unit mass at $\lambda=0$.

Example 2. The function $\varphi(\zeta)=\mathrm{Log}\,\zeta$; the logarithm being determined in such a way that it is real on the positive half-axis. The imaginary part $V(\zeta)$ is then bounded by π in the upper half-plane and $V(\lambda+i\eta)$ converges with decreasing η to π whenever $\lambda<0$. The ratio $\varphi(i\eta)/i\eta$ $=(\log\eta+i\pi/2)/i\eta$ converges to 0 with increasing η, whence $\alpha=0$. We have $\beta=\mathrm{Re}[\varphi(i)]=0$ and therefore

$$\mathrm{Log}\,\zeta=\int_{-\infty}^{0}\left[\frac{1}{\lambda-\zeta}-\frac{\lambda}{\lambda^2+1}\right]d\lambda\,.$$

Our result may be checked, since it is possible to compute the integral explicitly.

Example 3. The function $\varphi(\zeta)=\sqrt{\zeta}$: where that determination of the square root is taken which is positive on the right half-axis. Here, as before, $\alpha=0$ since $\varphi(i\eta)/i\eta$ converges to 0; while $\beta=\mathrm{Re}[\sqrt{i}]=1/\sqrt{2}$. The imaginary part $V(\lambda+i\eta)$ converges with diminishing positive η to 0 for $\lambda>0$ and to $\sqrt{|\lambda|}$ for $\lambda<0$. The measure $d\mu(\lambda)$ has no point mass at $\lambda=0$, since the function is positive on the right half-axis. Thus

$$\sqrt{\zeta}=\frac{1}{\sqrt{2}}+\int_{-\infty}^{0}\left[\frac{1}{\lambda-\zeta}-\frac{\lambda}{\lambda^2+1}\right]\frac{\sqrt{\lambda}}{\pi}\,d\lambda\,.$$

We have earlier remarked that the Pick class is closed under composition. Accordingly the function $\log(\varphi(\zeta))$ is in P whenever $\varphi(\zeta)$ is. The imaginary part of the logarithm is bounded, and therefore the canonical representation of $\log\varphi(\zeta)$ involves a measure $d\mu$ which is absolutely continuous relative to Lebesgue measure. Indeed, if $\log\varphi(\zeta)$ $=U+iV$, we have $V(\zeta)\leqq\pi$ in the upper half-plane, whence, for the measure, $d\mu(\lambda)=f(\lambda)\,d\lambda$ with $0\leqq f(\lambda)\leqq 1$. This gives rise to an exponential representation of $\varphi(\zeta)$:

$$\varphi(\zeta)=\exp\left[\sigma+\int\left[\frac{1}{x-\zeta}-\frac{x}{x^2+1}\right]f(x)\,dx\right] \tag{6}$$

where $\sigma=\mathrm{Re}[\log\varphi(i)]$. No coefficient of the form α occurs in (6), since the logarithm does not grow sufficiently fast at infinity. It is clear that any measurable $f(x)$ satisfying $0\leqq f(x)\leqq 1$ and any real constant σ give rise, via (6), to a function in P. For certain problems it is convenient to make use of the exponential representation rather than the canonical one. Here we remark only that if $\varphi(\zeta)$ in P is meromorphic and real on the real axis, then $\varphi(\zeta)$ is monotone increasing between its poles. Since the poles all have negative residues, $\varphi(\zeta)$ assumes all real values between any consecutive poles, and in particular, has one and only one zero between any two such poles. The function $\log\varphi(\zeta)$ can be continued analytically into the lower half-plane whenever $\varphi(\zeta)$ is not zero; the continuation is by reflection when it is over intervals where $\varphi(\zeta)>0$. If x is a point where $\varphi(x)$ is negative, the numbers $V(x+i\eta) = \operatorname{Im}[\log\varphi(x+i\eta)]$ converge with decreasing η to π, whence the function $f(x)$ in (6) is $+1$ at such an x, and equals 0 where $\varphi(x)>0$. It follows that $f(x)$ in (6) is simply the characteristic function of the set upon which $\varphi(x)$ is negative. As an example, we take $\varphi(\zeta)=\tan\zeta$; we have $\operatorname{Re}[\operatorname{Log}\tan i]=0$ and therefore

$$\tan\zeta=\exp\left[\operatorname{Log}\tanh 1+\int\left[\frac{1}{x-\zeta}-\frac{x}{x^2+1}\right]f(x)\,dx\right]$$

where $f(x)$ is the characteristic function of the set of x for which $\tan x<0$. Since the tangent is in P, so is the function $\varphi(\zeta)=-1/\tan\zeta=-\cot\zeta$, and for this function $\beta=\operatorname{Re}[\varphi(i)]=0$, as well as $\alpha=0$, since $\varphi(i\eta)$ is bounded for large η. The function is meromorphic, with poles at the zeros of the sine, and since the cotangent is the logarithmic derivative of the sine, the residues at those poles are integers. Thus we have the canonical representation

$$-\cot\zeta=\sum\left[\frac{1}{n\pi-\zeta}-\frac{n\pi}{n^2\pi^2+1}\right]$$

the summation being taken over all integers. Adding $1/\zeta$ to each side we get

$$\frac{1}{\zeta}-\cot\zeta=\sum_{n=1}^{\infty}\left[\frac{1}{n\pi-\zeta}-\frac{1}{n\pi+\zeta}\right]=\sum_{n=1}^{\infty}\frac{2\zeta}{n^2\pi^2-\zeta^2}$$

and therefore

$$\frac{d}{d\zeta}\left(\log\frac{\sin\zeta}{\zeta}\right)=\sum_{n=1}^{\infty}\frac{-2\zeta}{n^2\pi^2-\zeta^2}.$$

The series on the right converges absolutely and uniformly on any compact set bounded away from the integers; we may integrate along any path from 0 to z which does not pass through such a point; since the function $\sin\zeta/\zeta$ is $+1$ at the origin, its logarithm vanishes, whence

$$\log\frac{\sin z}{z}=\sum_{n=1}^{\infty}\int_0^z\frac{-2\zeta}{n^2\pi^2-\zeta^2}\,d\zeta=\sum_{n=1}^{\infty}\log(n^2\pi^2-\zeta^2)\Big|_0^z=\sum_{n=1}^{\infty}\log\left(\frac{n^2\pi^2-z^2}{n^2\pi^2}\right).$$

Taking exponentials, we finally obtain

$$\frac{\sin z}{z}=\prod_1^{\infty}\left(1-\frac{z^2}{n^2\pi^2}\right).$$

3. The Gamma Function

A similar argument can be used to study the Gamma function; we seek a meromorphic function $\Gamma(\zeta)$ having the following three properties:

(i) $\Gamma(\zeta)$ is real and regular on the right half-axis;

(ii) $\Gamma(1)=1$;

(iii) $\zeta\Gamma(\zeta)=\Gamma(\zeta+1)$.

If we differentiate the functional equation (iii) we have

$$\Gamma(\zeta)+\zeta\Gamma'(\zeta)=\Gamma'(\zeta+1)$$

whence, dividing by $\Gamma(\zeta+1)$, we obtain

$$\frac{\Gamma'(\zeta)}{\Gamma(\zeta)}+\frac{1}{\zeta}=\frac{\Gamma'(\zeta+1)}{\Gamma(\zeta+1)}.$$

Now put $\varphi(\zeta)=(d/d\zeta)\operatorname{Log}\Gamma(\zeta)=\Gamma'(\zeta)/\Gamma(\zeta)$ to obtain

$$\varphi(\zeta+1)-1/\zeta=\varphi(\zeta).$$

We look for solutions to this functional equation in the Pick class. We first remark that if $\varphi(\zeta)$ is a Pick function, so also is $\varphi(\zeta+1)$ since the class P is closed under composition, $\zeta+1$ being surely a Pick function. Thus all three functions occurring in the functional equation are in P. Moreover, if μ is the measure in the canonical representation (1) for $\varphi(\zeta)$ the measure corresponding to $\varphi(\zeta+1)$ is merely μ translated one unit to the left. Let us write that measure μ_1. Since both sides of the equation are Pick functions, and since $-1/\zeta$ is associated with the unit mass at the origin we have

$$\mu_1+\delta=\mu$$

where the δ measure is the unit mass at $\lambda=0$. Accordingly, the measure μ is unchanged if it is translated one unit to the left and a unit mass at the origin added in. It is evident that a solution to this equation is given by the measure μ_0 which puts a unit mass at every non-positive integer. Note that the function $(\lambda^2+1)^{-1}$ is μ_0 integrable, so that this measure is in fact associated with a Pick function. Moreover, if ν were another solution to the equation, i.e.

$$\nu_1+\delta=\nu$$

we would have $(\mu_0-\nu)_1=\mu_0-\nu$ and the difference would be a periodic measure with period 1. Hence the most general solution to our equation is of the form $\mu_0+\omega$ where ω is periodic of period 1. Now condition (1) which requires $\Gamma(\zeta)$ to be regular on the right half-axis makes it necessary for $\varphi(\zeta)$ to be in the class $P(0,\infty)$ and therefore for the measure to have no periodic part. Accordingly, $\mu=\mu_0$ and

$$\varphi(\zeta)=\alpha\zeta+\beta+\sum_{n=0}^{\infty}\left[\frac{1}{-n-\zeta}+\frac{n}{n^2+1}\right].$$

We can now write $\varphi(\zeta+1)-1/\zeta$ in the form

$$\alpha\zeta+\alpha+\beta-1/\zeta+\sum_{n=0}^{\infty}\left[\frac{1}{-n-1-\zeta}+\frac{n}{n^2+1}\right]$$

or better

$$\alpha\zeta+\alpha+\beta+\sum_{k=0}^{\infty}\left[\frac{1}{-k-\zeta}+\frac{k}{k^2+1}\right]-\sum_{k=1}^{\infty}\left[\frac{k}{k^2+1}-\frac{k-1}{(k-1)^2+1}\right].$$

The final sum vanishes and so from the functional equation for $\varphi(\zeta)$ we infer that $\beta=\alpha+\beta$, whence $\alpha=0$. We have now obtained a Pick function satisfying the functional equation. Note that this function is determined up to the value of β by the equation and the condition that $\varphi(\zeta)$ be regular far out on the right half-axis.

Consider the Pick function

$$\psi(\zeta)=\varphi(\zeta)+1/\zeta=\frac{d}{d\zeta}\operatorname{Log}(\zeta\Gamma(\zeta))=\beta+\sum_{n=1}^{\infty}\left[\frac{1}{-n-\zeta}+\frac{n}{n^2+1}\right].$$

Let $-C=\psi(0)$ and write $\psi(\zeta)=\psi(0)+\psi(\zeta)-\psi(0)$ in the form

$$\psi(\zeta)=-C+\sum_{n=1}^{\infty}\left[\frac{1}{n}-\frac{1}{n+\zeta}\right].$$

The series converges uniformly on compact subsets bounded away from the negative integers, a circumstance which enables us to interchange the summation and integration in the calculation below. Since $\zeta\Gamma(\zeta)$ is regular at the origin and equal to $+1$ there, its logarithm vanishes and we have

$$\operatorname{Log}(\zeta\Gamma(\zeta))=\int_0^{\zeta}\psi(z)\,dz$$

where we integrate along a line segment from the origin to ζ. Accordingly, at least for values of ζ that are not negative

$$\begin{aligned}\operatorname{Log}(\zeta\Gamma(\zeta))&=-C\zeta+\sum_{n=1}^{\infty}\int_0^{\zeta}\left[\frac{1}{n}-\frac{1}{n+z}\right]dz\\&=-C\zeta+\sum_{n=1}^{\infty}\left[\frac{\zeta}{n}-\operatorname{Log}\left(1+\frac{\zeta}{n}\right)\right].\end{aligned}$$

From the condition (ii) that fixes the value of Gamma at $+1$ we infer that $\operatorname{Log}(\zeta\Gamma(\zeta))$ vanishes at $\zeta=1$ and therefore that

$$-C+\sum_{n=1}^{\infty}\left[\frac{1}{n}-\operatorname{Log}\frac{n+1}{n}\right]=0$$

from which we readily infer that

$$C=\lim_{N\to\infty}\sum_{n=1}^{N}\frac{1}{n}-\operatorname{Log}(N+1)$$

thus, C is Euler's constant. Taking exponentials, we find

$$\zeta\Gamma(\zeta)=e^{-C\zeta}\prod_{n=1}^{\infty}\frac{e^{\zeta/n}}{1+\zeta/n}.$$

In this way we have obtained a meromorphic function $\Gamma(\zeta)$ which has the required properties. The fact that the series expansion for $\operatorname{Log}(\zeta\Gamma(\zeta))$ converges uniformly on compact subsets of the half-plane makes it clear that the resulting infinite product is convergent. The Gamma function now appears as a meromorphic function with poles at the non-negative integers. We adjusted the constant C to make sure that the function $\operatorname{Log}(\zeta\Gamma(\zeta))$ vanished at both $\zeta=0$ and $\zeta=1$, and so conditions (i) and (ii) are satisfied. To verify the functional equation, we write

$$L(z)=\operatorname{Log}(z\Gamma(z))=\int_0^z\psi(\zeta)\,d\zeta$$

and note that (iii) is equivalent to

$$L(z)=L(z+1)-\operatorname{Log}(z+1)$$

or

$$\int_0^z\psi(\zeta)\,d\zeta=\int_0^{z+1}\psi(\zeta)\,d\zeta-\int_1^{z+1}\frac{d\zeta}{\zeta}.$$

If we make use now of the fact that $\int_0^1\psi(\zeta)\,d\zeta=0$, the first integral on the right may be written $\int_1^{z+1}\psi(\zeta)\,d\zeta=\int_0^z\psi(\zeta+1)\,d\zeta$ and we therefore have to establish the functional equation

$$\int_0^z\psi(\zeta)\,d\zeta=\int_0^z\psi(\zeta+1)\,d\zeta-\int_0^z\frac{d\zeta}{\zeta+1}.$$

This equation, however, is an immediate consequence of the functional equation for $\varphi(\zeta)$.

4. Convergence of Pick Functions

There is a natural topology for the class of Pick functions, namely the topology of uniform convergence on compact subsets of the upper half-plane. It is obvious that under this topology P is a complete metric space. It is important to establish a certain compactness property for subsets of P.

Lemma 3. *Let $\zeta_0=\xi_0+i\eta_0$ be a point in the upper half-plane and $\mathscr{I}$ an infinite family of functions in P uniformly bounded at ζ_0, i.e. $|\varphi(\zeta_0)|\leqq M$ for every φ in $\mathscr{I}$; then there exists a sequence in $\mathscr{I}$ which is convergent.*

Proof. Let C be an arbitrary compact subset of the upper half-plane; we shall show that the family $\mathscr{I}$ is uniformly bounded on C and that the functions in the family are equicontinuous there. More exactly, we shall show that the functions in $\mathscr{I}$ are uniformly Lipschitzian on C. From the Ascoli-Arzela theorem, then, it will follow that there exists a convergent sequence in $\mathscr{I}$. There is no loss of generality in our supposing that ζ_0 belongs to C.

From the inequality

$$V(\zeta_0)=\alpha\eta_0+\int\frac{\eta_0\,d\mu(\lambda)}{|\lambda-\zeta_0|^2}\leqq M$$

we deduce that for any function $\varphi(\zeta)$ in the family

$$\alpha\leqq\frac{M}{\eta_0}\quad\text{and}\quad\int\frac{d\mu(\lambda)}{|\lambda-\zeta_0|^2}\leqq\frac{M}{\eta_0}.$$

Again, for any such function and any two points ζ_1 and ζ_2 in C we consider the difference quotient and find that

$$\left|\frac{\varphi(\zeta_1)-\varphi(\zeta_2)}{\zeta_1-\zeta_2}\right|\leqq\alpha+\int\frac{d\mu(\lambda)}{|\lambda-\zeta_1||\lambda-\zeta_2|}.$$

Now there exists a constant m, depending on the compact set C, such that if ζ varies over C and λ over the real axis, then

$$\frac{1}{m}\leqq\frac{|\lambda-\zeta|}{|\lambda-\zeta_0|}\leqq m.$$

The difference quotient is therefore bounded by

$$\alpha+\int\frac{m\,d\mu(\lambda)}{|\lambda-\zeta_0|^2}\leqq\frac{M(1+m)}{\eta_0}.$$

Virtually the same argument serves to establish the following results.

Lemma 4. *Let $\mathscr{I}$ be an infinite family of functions in $P(a, b)$ uniformly bounded on some subinterval of that interval: then there exists a sequence in $\mathscr{I}$ which converges in P to a limit in $P(a, b)$ and which also converges to that limit uniformly on compact subsets of (a, b).*

Lemma 5. *An infinite family of functions, positive and harmonic in the upper half-plane contains a sequence converging uniformly on compact subsets of the half-plane if it is uniformly bounded at some point. A sequence of such functions, unbounded at some fixed point, contains a subsequence which converges to infinity, i.e., the reciprocals converge to 0 uniformly on compact sets.*

Chapter III. Pick Matrices and Loewner Determinants

Let Z be a set of l points $\zeta_1, \zeta_2, \dots, \zeta_l$ in the upper half-plane and $\varphi(\zeta)$ a function defined for the upper half-plane. Form the Pick matrix associated with Z and $\varphi(\zeta)$:

$$K_{ij} = \frac{\varphi(\zeta_i) - \overline{\varphi(\zeta_j)}}{\zeta_i - \overline{\zeta}_j}$$

which is obviously symmetric and of order l.

Theorem I. *If $\varphi(\zeta)$ is a Pick function, the corresponding Pick matrix K is a positive matrix. K has the eigenvalue 0 with multiplicity k if and only if $\varphi(\zeta)$ is rational, of degree $l-k$ and real on the real axis. In this case the function is uniquely determined by the data, i.e. the set Z and the values taken by $\varphi(\zeta)$ on it.*

Proof. We make use of the canonical representation of $\varphi(\zeta)$ to compute

$$K_{ij} = \alpha + \int \frac{1}{\lambda - \zeta_i} \overline{\frac{1}{\lambda - \zeta_j}} \, d\mu(\lambda)$$

and therefore to write the quadratic form

$$\sum \sum K_{ij} z_i \overline{z}_j = \alpha \left|\sum z_i\right|^2 + \int \left| \sum \frac{z_i}{\lambda - \zeta_i} \right|^2 d\mu(\lambda) \geqq 0 .$$

Evidently K is a positive matrix. If 0 is an eigenvalue, there exists a non-trivial choice of the z_i's making the quadratic form vanish. The corresponding rational function of λ

$$F(\lambda) = \sum \frac{z_i}{\lambda - \zeta_i}$$

has at most $l-1$ finite zeros, since it vanishes at infinity and is of degree at most l. Clearly the measure μ must be concentrated on that finite set and it follows that $\varphi(\zeta)$ is real on the real axis and rational of degree at most l. The degree is in fact smaller than l, for if $\alpha = 0$, $\varphi(\zeta)$ has at most $l-1$ poles, since it is regular at infinity. However, if $\alpha \neq 0$, we have $\sum z_i = 0$, from which it follows that $F(\lambda)$ above has a double zero at infinity, and therefore at most $l-2$ zeros on the finite real axis. Thus $\varphi(\zeta)$ is again of degree at most $l-1$.

On the other hand, if we suppose that $\varphi(\zeta)$ is rational and real on the real axis with degree $l-k$ and with $\alpha=0$ then the Pick matrix is a Gram's matrix:

$$K_{ij} = (f_i, f_j) \quad \text{where } f_i(\lambda) = (\lambda - \zeta_i)^{-1}$$

the inner product being taken in the L^2-space associated with the measure μ. The dimension of this space is $l-k$ since the measure consists only of that many point masses, and the system of functions $f_i(\lambda)$ span the space. Accordingly the matrix K has a k-dimensional null space.

The proof of the theorem is almost complete, except for the special case where $\alpha \neq 0$. If $\varphi(\zeta)$ is of degree $l-k$ with a positive α, we pass to the function

$$\psi(\zeta) = \frac{-1}{\varphi(\zeta)}$$

which is also a Pick function, rational of degree $l-k$. For this new function the corresponding value of α is 0. $\psi(\zeta)$ is associated with the Pick matrix

$$K'_{ij} = \frac{1}{\varphi(\zeta_i)\overline{\varphi(\zeta_j)}} \frac{\varphi(\zeta_i) - \overline{\varphi(\zeta_j)}}{\zeta_i - \bar{\zeta}_j}$$

and so

$$K_{ij} = \varphi(\zeta_i) K'_{ij} \overline{\varphi(\zeta_j)}$$

which may be written $K = F K F^*$ where F is a diagonal matrix with entries $\varphi(\zeta_i)$. Since none of the values of $\varphi(\zeta)$ vanishes, F is an invertible matrix, and K and K' have null spaces of the same dimension. Our theorem therefore holds in this case too.

The fact that $\varphi(\zeta)$ is uniquely determined by the data when the matrix K is singular is a consequence of the fact that $\varphi(\zeta)$ is real on the real axis, and so satisfies the l additional equations $\varphi(\bar{\zeta}_i) = \overline{\varphi(\zeta_i)}$ by virtue of the Reflection Principle. Thus $\varphi(\zeta)$, of degree at most $l-1$, is determined at $2l$ points. It follows that it is determined uniquely.

It is interesting to consider next the study of a complex-valued function $\varphi(\zeta)$ in the upper half-plane subject to only one condition: that for any two points in the half-plane, the 2 by 2 Pick matrix K is a positive matrix. One deduces immediately that the function has a positive imaginary part, since the diagonal elements of the matrix

$$K_{jj} = \frac{2i \operatorname{Im}[\varphi(\zeta_j)]}{2i\eta_j}$$

are positive.

A closer analysis enables us to infer that the function is continuous. For this purpose, the assertion that the determinant of the matrix is positive leads to the inequality

$$\frac{|\varphi(\zeta_1)-\overline{\varphi(\zeta_2)}|^2}{|\varphi(\zeta_1)-\overline{\varphi(\zeta_1)}|\,|\varphi(\zeta_2)-\overline{\varphi(\zeta_2)}|} \leq \frac{|\zeta_1-\bar{\zeta}_2|^2}{|\zeta_1-\bar{\zeta}_1|\,|\zeta_2-\bar{\zeta}_2|}.$$

Call the ratio on the right $R(\zeta_1, \zeta_2)$ and note that it equals 1 if and only if $\zeta_1=\zeta_2$, otherwise it is larger. Our inequality then reads

$$R(\varphi(\zeta_1), \varphi(\zeta_2)) \leqq R(\zeta_1, \zeta_2).$$

If ζ_2 approaches ζ_1, then $R(\zeta_1, \zeta_2)$ converges to 1, whence $R(\varphi(\zeta_1), \varphi(\zeta_2))$ also converges to 1. This implies that the number $\varphi(\zeta_2)$ converges to $\varphi(\zeta_1)$ since the family of sets of all w in the half-plane for which $R(\varphi(\zeta_1), w) < 1+\varepsilon$ as ε approaches 0 form a decreasing family of circular neighborhoods of $\varphi(\zeta_1)$. Thus $\varphi(\zeta)$ is continuous in the half-plane.

A closer study of the function will not provide any great regularity properties for $\varphi(\zeta)$, and certainly the function need not be analytic, as the example

$$\varphi(\zeta) = \frac{x}{2} + iy$$

shows. However, if we require that all Pick matrices of order 3 should be positive, the function $\varphi(\zeta)$ is necessarily analytic, as the following remarkable theorem due to Hindmarsh shows.

Theorem II. *Let $\varphi(\zeta)$ be defined in the upper half-plane and satisfy the condition that for every three points z_1, z_2, z_3 in the half-plane the Pick matrix*

$$K(z_i, z_j) = \frac{\varphi(z_i)-\overline{\varphi(z_j)}}{z_i-\bar{z}_j}$$

is a positive matrix. Then $\varphi(\zeta)$ is a Pick function.

Proof. We have already seen that the function must have a positive imaginary part in the upper half-plane and that it is continuous. All that needs to be shown is the analyticity of the function, and it is enough for us to show this for any disk in the half-plane, bounded away from the real axis. We first prove the theorem under the special hypothesis that the function is in the class C^2.

Choose a point z in the half-plane and a small positive h and form the matrix

$$K = \begin{bmatrix} K(z,z) & K(z,z+h) & K(z,z+ih) \\ K(z+h,z) & K(z+h,z+h) & K(z+h,z+ih) \\ K(z+ih,z) & K(z+ih,z+h) & K(z+ih,z+ih) \end{bmatrix}.$$

By hypothesis, K is a positive matrix. Now form the positive matrix $A^* K A$ where

$$A = \begin{bmatrix} 1 & -1/h & -1/h \\ 0 & 1/h & 0 \\ 0 & 0 & 1/h \end{bmatrix}.$$

The elements of $A^* K A$ are difference quotients of the function $K(z, \zeta)$

$$K(z, \zeta) = \frac{\varphi(z) - \overline{\varphi(\zeta)}}{z - \overline{\zeta}}$$

and as h tends to 0 these elements tend to certain derivatives of that function which is C^2 by hypothesis. We obtain the positive matrix

$$N = \begin{bmatrix} K & K_\xi & K_\eta \\ K_x & K_{x\xi} & K_{x\eta} \\ K_y & K_{y\xi} & K_{y\eta} \end{bmatrix}.$$

Here the element K_ξ is the derivative with respect to ξ of $K(x+iy, \xi+i\eta)$ evaluated at $\zeta = z$. Similarly K_y is the derivative with respect to y of that function, evaluated at the same point.

Since we are trying to show a function analytic, it is convenient to pass to the derivatives in terms of the variables $z, \overline{z}, \zeta$ and $\overline{\zeta}$, since then the Cauchy-Riemann equations take the simple form $(\partial/\partial \overline{z})\varphi(z) = 0$. We will have

$$\frac{\partial}{\partial z} = \frac{1}{2}\left(\frac{\partial}{\partial x} + \frac{1}{i}\frac{\partial}{\partial y}\right) \quad \text{and} \quad \frac{\partial}{\partial \overline{z}} = \frac{1}{2}\left(\frac{\partial}{\partial x} - \frac{1}{i}\frac{\partial}{\partial y}\right).$$

We therefore write

$$B = \begin{bmatrix} 1 & 0 & 0 \\ 0 & 1/2 & -i/2 \\ 0 & 1/2 & i/2 \end{bmatrix}$$

and form the positive matrix

$$B N B^* = \begin{bmatrix} K & K_{\overline{\zeta}} & K_\zeta \\ K_z & K_{z\overline{\zeta}} & K_{z\zeta} \\ K_{\overline{z}} & K_{\overline{z}\overline{\zeta}} & K_{\overline{z}\zeta} \end{bmatrix}.$$

We will show that the element in the lower right hand corner of the matrix is 0; this will imply, since the matrix is positive, that everything in the bottom row vanishes, in particular $K_{\overline{z}}(z, z)$, and this, in turn, will show that $\varphi(z)$ satisfies the Cauchy-Riemann equations. Now

$$K_{\overline{z}}(z, \zeta) = \frac{\partial}{\partial \overline{z}} \frac{\varphi(z) - \overline{\varphi(\zeta)}}{z - \overline{\zeta}} = \frac{1}{z - \overline{\zeta}} \frac{\partial}{\partial \overline{z}} \varphi(z).$$

Hence
$$K_{\bar{z}\zeta}(z,\zeta)=0\,.$$
We infer that $K_{\bar{z}}(z,z)=(1/2iy)/(\partial/\partial\bar{z})\,\varphi(z)=0$.

To complete the proof we consider the more general case where $\varphi(z)$ is continuous, but not necessarily C^2. However, the regularizations of that function will also satisfy the three point property stated in the hypothesis of the theorem, and these will be C^2-functions and therefore analytic. Since the regularizations converge uniformly on compact subsets of the half-plane to the continuous function $\varphi(z)$, it follows that $\varphi(z)$ is itself analytic. The proof is complete.

We now turn to theorems analogous to the principal theorem of this chapter, but which refer to the class $P(a,b)$.

Theorem III. *Let $\varphi(\zeta)$ belong to $P(a,b)$ and $\lambda_1,\lambda_2,\ldots,\lambda_l$ be l distinct points of the open interval (a,b); then the Pick matrix formed with the divided differences*
$$K_{ij}=[\lambda_i,\lambda_j]$$
is a positive matrix. K has the eigenvalue 0 with multiplicity k if and only if $\varphi(\zeta)$ is rational and of degree $l-k$. In this case the function is uniquely determined by the data, i.e., the values taken by $\varphi(\zeta)$ and its derivative at the points λ_i.

We do not prove the theorem just stated, since its proof is remarkably similar to the proof given earlier; the canonical representation of $\varphi(\zeta)$ makes it immediately evident that K_{ij} is a Gram's matrix, at least when $\alpha=0$. Only a little thought is needed to calculate the degree of the rational $\varphi(\zeta)$ when K is a singular matrix.

Another theorem is also easy to prove.

Theorem IV. *Let $\varphi(\zeta)$ belong to $P(a,b)$ and let c be a point of that interval. If*
$$a_k=\frac{\varphi^{(k)}(c)}{k!}$$
is the k-th Taylor coefficient for $\varphi(\zeta)$ at c, then the matrix
$$A=\begin{bmatrix} a_1 & a_2 & \ldots & a_l \\ a_2 & a_3 & \ldots & \\ \vdots & \vdots & & \vdots \\ a_l & & \ldots & a_{2l-1} \end{bmatrix}$$
is a positive matrix. A has the eigenvalue 0 with multiplicity k if and only if $\varphi(\zeta)$ is rational and of degree $l-k$. In this case the function $\varphi(\zeta)$ is uniquely determined by the data, i.e., the value of the function and its first $2l-1$ derivatives at c.

Another theorem concerning functions in $P(a,b)$ is more complicated and refers to determinants rather than matrices. Here we suppose that $\xi_1, \xi_2, \ldots, \xi_l$ are l distinct points of the interval (a,b); similarly, we suppose that $\eta_1, \eta_2, \ldots, \eta_l$ are also l distinct points of that interval, although we make no hypothesis that the sets ξ_i and η_j are disjoint. We form the Loewner matrix from the divided differences of a function $\varphi(\zeta)$ in $P(a,b)$:

$$L_{ij} = [\xi_i, \eta_j].$$

Evidently the matrix elements L_{ij} are non-negative.

Theorem V. *If $\varphi(\zeta)$ belongs to $P(a,b)$, then the Loewner determinant* $\det L$ *vanishes if and only if $\varphi(\zeta)$ is rational, of degree at most $l-1$. If the sequences ξ_i and η_j are both monotone increasing, then* $\det L \geqq 0$.

Proof. We first prove the theorem under the special hypothesis that the coefficient $\alpha = 0$ in the canonical integral representation for $\varphi(\zeta)$. Using that representation we easily find

$$L_{ij} = \int \frac{d\mu(\lambda)}{(\lambda - \xi_i)(\lambda - \eta_j)}$$

and this may be written $L_{ij} = (f_i, g_j)$, where $f_i(\lambda) = (\lambda - \xi_i)^{-1}$ and $g_j(\lambda) = (\lambda - \eta_j)^{-1}$. The functions $f_i(\lambda)$ and $g_j(\lambda)$ are evidently in the L^2-space associated with the measure μ, which has no mass in the interval (a,b) where the ξ's and η's are.

If $\varphi(\zeta)$ is rational, of degree at most $l-1$, then the measure μ consists of point masses at the poles of the function, and there are at most $l-1$ of them. Consequently, the dimension of the Hilbert space $L^2(\mu)$ is at most $l-1$ and there must exist a linear dependence between the functions $f_i(\lambda)$. Accordingly, the determinant of L is 0.

Conversely, if the determinant vanishes, there exists a linear combination

$$f^*(\lambda) = \sum z_i f_i(\lambda)$$

which is orthogonal to every $g_j(\lambda)$. Let $Q(\lambda) = \prod_{i=1}^{l} (\lambda - \xi_i)$; we can then write $f^*(\lambda) = p(\lambda)/Q(\lambda)$, where $p(\lambda)$ is a polynomial of degree at most $l-1$. Now set $R(\lambda) = \prod_{j=1}^{l} (\lambda - \eta_j)$ and form the function $g^*(\lambda) = p(\lambda)/R(\lambda)$. The function $g^*(\lambda)$ is a linear combination of the $g_j(\lambda)$, and so $(f^*, g^*) = 0$. This may be written

$$\int \frac{p(\lambda)^2}{Q(\lambda)R(\lambda)} d\mu(\lambda) = 0$$

and since the denominator $Q(\lambda)R(\lambda)$ is positive on the support of μ we infer that the measure is concentrated at the zeros of $p(\lambda)$. Hence $\varphi(\zeta)$ is rational, of degree at most $l-1$.

If we next suppose that $\varphi(\zeta)$ is not rational of degree at most $l-1$ then the determinant is not zero. Assuming that the sequences ξ_i and η_j are monotone, we form

$$\eta_j(t)=t\xi_j+(1-t)\eta_j$$

as t runs over the unit interval. The Loewner determinant formed with the ξ_i and the $\eta_j(t)$ will be a continuous, non-vanishing function of t. For $t=0$ the η_j coincide with the ξ_j and the matrix of divided differences is evidently a positive matrix, hence with a positive determinant. For $t=1$ the determinant is therefore also positive.

It remains to extend the proof of the theorem to the case when $\alpha\neq 0$. For this purpose, choose a slightly smaller interval (a', b') containing the points ξ_i and η_j and select $\lambda>0$ so that $\lambda\geqq\varphi(b')$. Now form the function

$$\psi(\zeta)=\frac{\lambda\varphi(\zeta)+1}{\lambda-\varphi(\zeta)};$$

as the composition of two Pick functions, $\psi(\zeta)$ is also a Pick function. Evidently this function is regular in the interval (a', b'), so $\psi(\zeta)$ is in $P(a', b')$. Let L' be the Loewner matrix corresponding to the function $\psi(\zeta)$ and the points ξ_i and η_j; an easy computation shows that

$$L'_{ij}=\frac{(\lambda^2+1)}{(\lambda-\varphi(\xi_i))(\lambda-\varphi(\eta_j))}L_{ij}.$$

Accordingly, $\det L'=C\det L$, where C is the positive constant

$$(\lambda^2+1)^l\prod_{i=1}^{l}(\lambda-\varphi(\xi_i))^{-1}(\lambda-\varphi(\eta_i))^{-1}.$$

Accordingly, the one determinant is positive when the other is, while $\psi(\zeta)$ is rational, of degree k, if and only if $\varphi(\zeta)$ is. Since $\psi(\zeta)$ is a function for which $\alpha=0$, the proof is complete.

The results of the previous theorem may be extended to cover the case of the determinant of the extended Loewner matrix. Here, as before, we suppose that $\varphi(\zeta)$ is in the class $P(a, b)$ and that ξ_i and η_j are each two monotone sequences of l distinct points. We then form the extended Loewner matrix:

$$L_e=\begin{bmatrix} [\xi_1,\eta_1] & [\xi_1,\eta_1,\eta_2] & [\xi_1,\eta_1,\eta_2,\eta_3] & & [\xi_1,\eta_1,\ldots,\eta_l] \\ [\xi_1,\xi_2,\eta_1] & [\xi_1,\xi_2,\eta_1,\eta_2] & & & \\ [\xi_1,\xi_2,\xi_3,\eta_1] & & & & \\ \vdots & & & & \\ [\xi_1,\xi_2,\ldots,\xi_l,\eta_1] & & & & [\xi_1,\xi_2,\ldots,\xi_l,\eta_1,\ldots,\eta_l] \end{bmatrix}$$

The extension of the results of the previous theorem to this matrix is almost immediate. We may first suppose that all the points ξ_i and η_j are distinct and make use of the recursive definition of the divided differences. The elements of the extended Loewner matrix are obtained from the original Loewner matrix by successive subtractions of rows from rows, and division by differences $\xi_i - \xi_j$, as well as a similar sequence of operations on the columns. For instance, our original Loewner determinant is equal to

$$\prod_{i>1} (\xi_i - \xi_1) \det \begin{bmatrix} [\xi_1, \eta_1] & [\xi_1, \eta_2] & \cdots [\xi_1, \eta_l] \\ [\xi_1, \xi_2, \eta_1] & [\xi_1, \xi_2, \eta_2] & \cdots [\xi_1, \xi_2, \eta_l] \\ \vdots & \vdots & \vdots \\ [\xi_1, \xi_l, \eta_1] & [\xi_1, \xi_l, \eta_2] & \cdots [\xi_1, \xi_l, \eta_l] \end{bmatrix}.$$

We finally obtain

$$\det L = \prod_{i>j} (\xi_i - \xi_j)(\eta_i - \eta_j) \det L_e .$$

It is now evident that $\det L_e$ is positive if the ξ_i and η_j form a monotone sequence. Since the divided differences are smooth functions of their arguments, and since the determinant depends continuously on the matrix elements we have proved another theorem, where we need not insist that the points ξ_i or η_j be distinct.

Theorem VI. *If the points ξ_i and η_j are monotone sequences in the interval (a, b) and if $\varphi(\zeta)$ belongs to $P(a, b)$, then the corresponding extended Loewner matrix has a non-negative determinant.*

Chapter IV. Fatou Theorems

In this chapter we investigate more closely the relation between a positive harmonic function $V(\zeta)$ in the upper half-plane and the measure μ occuring in its canonical representation. It is clear that the function is perfectly smooth, indeed harmonic, near any real point λ_0 which is at a positive distance from the support of the measure. There is also no loss of generality in our supposing that we study the function near the origin, that $\alpha=0$ and that the measure is supported by the interval $[-1,1]$.

Lemma 1. *If the function $V(\zeta)$ is bounded on some ray issuing from the origin then it is bounded on all other such rays. If $V(\zeta)$ becomes infinite as ζ approaches 0 along a ray, the function becomes infinite along any other ray.*

Proof. Suppose θ_1 and θ_2 are two values of the argument such that $V(re^{i\theta_1})$ is bounded as r approaches 0 but $V(re^{i\theta_2})$ is not bounded as r diminishes. From the system of positive harmonic functions

$$V_r(\zeta)=V(r\zeta)$$

we can extract a sequence converging to infinity at the point $\zeta_2=e^{i\theta_2}$. Since that sequence is uniformly bounded at the point $\zeta_1=e^{i\theta_1}$ there exists a further subsequence converging uniformly on compact subsets of the half-plane to a positive harmonic limit $V_0(\zeta)$ which is finite at ζ_2. This is a contradiction. If $V_r(e^{i\theta_1})$ converges to infinity the same argument shows that $\liminf V(re^{i\theta})$ is infinite for any other value of θ.

Lemma 2. *Let θ_1 and θ_2 be two values of the argument for which $V(\zeta)$ converges to a finite limit along the corresponding ray, i.e.,*

$$\lim_{r\to 0} V(re^{i\theta_1})=A_1 \quad \text{and} \quad \lim_{r\to 0} V(re^{i\theta_2})=A_2\,.$$

Then, uniformly over any closed subinterval of $(0,\pi)$

$$\lim_{r\to 0} V(re^{i\theta}) \quad \textit{exists and equals} \quad C\theta+D$$

where

$$C=\frac{A_1-A_2}{\theta_1-\theta_2} \quad \text{and } D=A_1-C\theta_1 .$$

Proof. As in the previous argument, from the family of positive harmonic functions $V_r(\zeta)=V(r\zeta)$ we extract a convergent subsequence which converges to a positive harmonic $V_0(\zeta)$. On the ray $z=re^{i\theta_1}$ the function $V_0(\zeta)$ is a constant and takes the value A_1; similarly the function takes the constant value A_2 on the ray with argument θ_2. We may suppose $0<\theta_1<\theta_2<\pi$ and pass to the function

$$W(\zeta)=V_0(e^{i\theta_1}\zeta^{\gamma})$$

where $\gamma=(\theta_2-\theta_1)/\pi$. $W(\zeta)$ is positive and harmonic in the upper half-plane and is even harmonic in the neighborhood of any real point $x\neq 0$. For $x>0$ $W(x)=A_1$, while the function equals A_2 for $x<0$. However, the function defined by the expression

$$(A_2-A_1)\theta/\pi+A_1=A_1+\frac{A_2-A_1}{\pi}\operatorname{Im}[\log z]$$

is also positive and harmonic in the upper half-plane and has exactly the same boundary values as $W(\zeta)$. It follows that both functions have the same measure in their canonical representation and therefore that they coincide. This makes $W(\zeta)$ linear in the angle θ and so $V_0(\zeta)$ also is for θ in the interval (θ_1, θ_2). However, the uniqueness of the harmonic continuation for $V_0(\zeta)$ guarantees that the function is linear in the angle throughout the upper half-plane. This essentially completes the proof of the lemma, the uniformity of the convergence of $V(re^{i\theta})$ to $V_0(e^{i\theta})$ on closed subintervals of $(0, \pi)$ being a consequence of the convergence of $V_r(\zeta)$ to $V_0(\zeta)$ on compact subsets of the half-plane.

We consider next the measure μ occuring in the canonical representation of $V(\zeta)$ and the corresponding monotone increasing function $\mu(\lambda)$ so normalized that $\mu(0)=0$ and $\mu(\lambda)=(\mu(\lambda+0)+\mu(\lambda-0))/2$. If $\mu(\lambda)$ has a derivative at the origin we write it $\mu'(0)$, while if $\theta(\lambda)$ is the ratio

$$\frac{\mu(\lambda)+\mu(-\lambda)}{2\lambda}$$

defined for $\lambda\neq 0$, the symmetric derivative of $\mu(\lambda)$ at the origin is the limit

$$D\mu(0)=\lim_{\lambda\to 0}\theta(\lambda)$$

if it exists. We will write $D_+\mu(0)$ and $D_-\mu(0)$ for the right and left hand derivatives of μ at the origin, and note that of the three quantities $D\mu(0)$, $D_+\mu(0)$ and $D_-\mu(0)$, if any two exist and are finite, then so is the third.

If any two exist and are equal, then $\mu'(0)$ exists and equals the common value.

We consider the function $V(\zeta)$ on the imaginary axis and write the canonical representation:

$$V(i\eta)=\int_{-\infty}^{+\infty}\frac{\eta}{\lambda^2+\eta^2}\,d\mu(\lambda)\geqq\frac{1}{\eta}\int_{-\eta}^{\eta}\frac{\eta^2}{\lambda^2+\eta^2}\,d\mu(\lambda)\geqq\frac{\mu(\eta)-\mu(-\eta)}{2\eta}=\theta(\eta).$$

We infer that if $V(\zeta)$ is bounded as ζ approaches the origin along the imaginary axis, then the function $\theta(\lambda)$ is bounded on the real axis. It is clear in any case that that function is even and positive, small and continuous for large $|\lambda|$. Integrating by parts we obtain

$$\begin{aligned}V(i\eta)&=\int_{-\infty}^{+\infty}\mu(\lambda)\frac{2\lambda\eta}{(\lambda^2+\eta^2)^2}\,d\lambda=\frac{1}{\eta}\int_{-\infty}^{+\infty}\mu(t\lambda)\frac{2t}{(t^2+1)^2}\,dt\\&=\int_0^{\infty}\theta(t\eta)\frac{4t^2}{(t^2+1)^2}\,dt\end{aligned}$$

and since $4t^2/(t^2+1)^2$ is an integrable function on the half-axis with integral π it is clear that the boundedness of $\theta(\lambda)$ implies the boundedness of $V(i\eta)$. Thus $V(i\eta)$ and $\theta(\lambda)$ are either both bounded or both unbounded. Moreover, if the symmetric derivative $D\mu(0)$ exists, even if it is infinite, the function $V(i\eta)$ converges with diminishing η to $D\mu(0)\pi$. This circumstance leads to a proof of the well-known Fatou Theorem.

Theorem I. *If the derivative $\mu'(0)$ exists, then $V(re^{i\theta})$ converges with diminishing r to $\mu'(0)\pi$ uniformly on closed subintervals of* $(0,\pi)$.

Proof. If $\mu'(0)$ is infinite then $V(i\eta)$ converges to infinity as η approaches 0 and from Lemma 1 it follows that $V(re^{i\theta})$ converges to infinity for all θ. If the derivative is finite we write

$$V(\zeta)=V_-(\zeta)+V_+(\zeta)$$

where $V_-(\zeta)$ correspond to a measure v_- which is the restriction of μ to the left half-axis; similarly the measure v_+ associated with $V_+(\zeta)$ is the restriction of μ to the right half-axis. The monotone function $v_-(\lambda)$ has a left hand and a right hand derivative at the origin: we will have $D_-v_-(0)=\mu'(0)$, $D_+v_-(0)=0$ and so $Dv_-(0)=\mu'(0)/2$. We pass to the function $W(\zeta)=V_-(\sqrt{\zeta})$ and notice that this function corresponds to a measure supported by the left half-axis of the form

$$\frac{1}{\pi}V_-(i\sqrt{|\lambda|})\,d\lambda$$

where $d\lambda$ is Lebesgue measure. Evidently the associated monotone function has a left hand derivative at the origin equal to $\mu'(0)/2$ since $V_-(i\eta)$ converges to a finite limit with diminishing η. It follows that a symmetric derivative exists since the right hand derivative is obviously 0. This means that $W(i\eta)$ converges to a finite limit as η approaches 0 and therefore that $V_-(\zeta)$ converges to a finite limit as ζ approaches 0 along the ray $\theta=\pi/4$. From Lemma 2 it now follows that $V_-(\zeta)$ approaches a finite limit along any ray and this limit has the form $\theta\mu'(0)$. The same argument, now applied to the function $V_+(\zeta)$, shows that this function converges to $(\pi-\theta)\mu'(0)$ as ζ approaches 0 along a ray. Hence, finally, along any ray $V(\zeta)$ converges to $\mu'(0)\pi$.

The Fatou theorem also has a converse.

Theorem II. *If $V_r(\zeta)=V(r\zeta)$ converges to a finite limit as the positive r approaches 0, then $\mu(\lambda)$ has a finite derivative at the origin.*

Proof. Let the limit in question be $A\geqq 0$ so that

$$V_r(\xi+i\eta)=\int\frac{r\eta}{(\lambda-r\xi)^2+r^2\eta^2}d\mu(\lambda)=\int\frac{\eta}{(\lambda-\xi)^2+\eta^2}\,\frac{1}{r}\,d\mu(r\lambda)$$

converges to A as r converges to 0. We consider the family of measures defined by

$$d\mu_r(\lambda)=\frac{1}{r}\,d\mu(r\lambda)$$

and note that

$$\int\frac{d\mu_r(\lambda)}{1+\lambda^2}=V(ri)$$

is uniformly bounded. We can therefore invoke Helly's Theorem for the family of measures

$$d\nu_r(\lambda)=(1+\lambda^2)^{-1}\,d\mu_r(\lambda)$$

considered on the one-point compactification of the real axis. Thus, passing if need be to a sequence of values of r converging to 0 we find a measure ν on the axis with the property that

$$\int F(\lambda)\,d\nu_r(\lambda)\quad\text{converges to}\quad\int F(\lambda)\,d\nu(\lambda)$$

for every function $F(\lambda)$ continuous on the compactified axis. Among such functions $F(\lambda)$ are the functions of the form

$$F(\lambda)=\frac{(1+\lambda^2)}{(\lambda-\xi)^2+\eta^2}$$

where $\eta > 0$, and so

$$V_r(\xi + i\eta) = \int \frac{\eta}{(\lambda-\xi)^2+\eta^2} \, d\mu_r(\lambda)$$

converges to

$$\int \frac{(1+\lambda^2)}{(\lambda-\xi)^2+\eta^2} \, d\nu(\lambda) = A \int \frac{\eta}{(\lambda-\xi)^2+\eta^2} \, \frac{1}{\pi} \, d\lambda .$$

Now from the uniqueness of the measure representing a positive harmonic function we deduce that the measure ν is uniquely determined and that it was not necessary to pass to a sequence of values of r converging to 0. The measures $d\mu_r$ have uniformly bounded mass on any bounded subinterval of the axis and these converge for such a subinterval to the measure $(A/\pi)\,d\lambda$ where, of course, $d\lambda$ is Lebesgue measure. An easy limiting argument then shows that if $F(\lambda)$ is the characteristic function of an interval (a,b) then $\int F(\lambda)\,d\mu_r(\lambda)$ converges to $(b-a)A/\pi$. However, we can compute the integral explicitly to obtain

$$\frac{\mu(rb)-\mu(ra)}{rb-ra}(b-a)$$

and therefore infer that the ratio

$$\frac{\mu(rb)-\mu(ra)}{rb-ra}$$

converges to A/π. Since the interval (a,b) was arbitrary, it follows that $\mu(\lambda)$, has a derivative at the origin. The proof is complete.

Concerning the symmetric derivative we obtain a similar theorem.

Theorem III. *The symmetric derivative $D\mu(0)$ exists and is finite if and only if $\lim_{\eta\to 0} V(i\eta)$ exists and is finite.*

Proof. We have already shown that the existence of a finite symmetric derivative guarantees that $V(i\eta)$ converges with diminishing η to $D\mu(0)\pi$. We shall therefore suppose that $V(i\eta)$ converges to a finite limit; that function will be bounded and so, by a remark made earlier, $\theta(\lambda)$ will be bounded on the real axis. We make a change of variables in the integral representing $V(i\eta)$ in terms of $\theta(\lambda)$ putting $t=e^x$ and $\eta = e^{-s}$ to obtain

$$V(ie^{-s}) = \int_{-\infty}^{+\infty} K(x) f(x-s)\, dx$$

where $K(x) = e^x/\cosh^2 x$ is a positive integrable function and $f(x) = \theta(e^x)$ is a bounded, nonnegative function, locally of bounded variation.

Moreover

$$\int_{-\infty}^{+\infty} K(x)\,dx = \int_{0}^{\infty} \frac{4t^2}{(t^2+1)^2}\,dt = \pi$$

and by hypothesis there exists a constant C so that the limit

$$\lim_{s\to\infty} \int_{-\infty}^{+\infty} K(x)\,f(x-s)\,dx$$

exists and equals πC. We shall first show that the Fourier transform of $K(x)$ has no real zero; this will permit us to apply the Wiener Trauberian theorem to infer that certain averages of the function $f(x)$ converge to C as x approaches $-\infty$. A somewhat more detailed analysis of the function $f(x)$ will then show that $f(x)$ itself converges to C at $-\infty$, and this will mean that $\theta(\lambda)$ converges to C as λ approaches 0, that is to say, that $D\mu(0)$ exists and equals C.

To show that the Fourier transform in question has no real zero we must show that for any real value of ξ the integral

$$\int_{-\infty}^{+\infty} e^{-ix\xi} \frac{e^x}{\cosh^2 x}\,dx$$

is not zero. We introduce the meromorphic function $F_\xi(z) = e^{(1-i\xi)z}/\cos^2(iz)$ and note that for real x $F_\xi(x+i\pi) = -e^{\pi\xi} F_\xi(x)$. The function has poles on the imaginary axis, and it is easy to see that if C is a circle of small radius about the point $i\pi/2$ then $\int_C F_\xi(z)\,dz$ is the same as the sum

$$\int_{-\infty}^{+\infty} F_\xi(x)\,dx + \int_{+\infty}^{-\infty} F_\xi(x+i\pi)\,dx$$

since $|F_\xi(x+iy)|$ diminishes rapidly for large $|x|$ uniformly in y. We have then

$$(1+e^{\pi\xi}) \int_{-\infty}^{+\infty} F_\xi(x)\,dx = \int_C F_\xi(z)\,dz$$

and after a suitable change of variable the integral becomes

$$e^{\xi\pi/2} \int \frac{e^{-(i+\xi)z}}{\sin^2 z}\,dz$$

the integration now being taken over a circle around the origin. Since $1/\sin^2 z = 1/z^2 + h(z)$ where $h(z)$ is analytic near the origin, we may substitute z^2 for $\sin^2 z$ in the integral and evaluate it by the Cauchy

Integral Formula to find that

$$(1+e^{\pi\xi})\int_{-\infty}^{+\infty} F_\xi(x)\,dx = e^{\xi\pi/2}\,2\pi(1-i\xi)$$

whence, finally

$$\int_{-\infty}^{+\infty} F_\xi(x)\,dx = \frac{\pi(1-i\xi)}{\cosh(\xi\pi/2)}$$

and this vanishes for no real ξ. From the Wiener Tauberian theorem we can now infer that for every function $H(z)$ integrable over the real axis the integral

$$\int_{-\infty}^{+\infty} H(x)\,f(x-s)\,dx$$

converges to $C\int_{-\infty}^{+\infty} H(x)\,dx$ as s approaches infinity. In particular, taking $H(x)=1/h$ for x in the interval $(0,h)$ and zero elsewhere we find that

$$\frac{1}{h}\int_{-s}^{-s+h} f(x)\,dx = f_h(-s)$$

converges to C as s increases. We want next to infer that $f(x)$ itself approaches C as x tends to $-\infty$ and this we can do since the function is relatively regular. The function $\lambda\theta(\lambda)$ is monotone non-decreasing for positive λ and so $e^x f(x)$ is non-decreasing as a function of x. For any positive t, then

$$e^{x+t} f(x+t) \geqq e^x f(x)$$

and therefore

$$\frac{f(x+t)-f(x)}{t} \geqq -f(x)\frac{1-e^{-t}}{t} \geqq -M$$

where M is a bound for $f(x)$. Suppose x_n is a sequence of points converging to $-\infty$ such that $f(x_n)\geqq C+2\varepsilon$ for some positive ε; the function then exceeds $C+\varepsilon$ on the interval $(x_n, x_n+\varepsilon/M)$ and so if $0<h<\varepsilon/M$ then $f_h(x_n)\geqq C+\varepsilon$ contradicting the fact that this function converges to C as x tends to $-\infty$. In a similar way we show that if $f(x_n)\leqq C-2\varepsilon$ for an appropriate sequence converging to $-\infty$ then the function is bounded by $C-\varepsilon$ over the intervals $(x_n-\varepsilon/M, x_n)$ so that for sufficiently small h $f_h(x_n-\varepsilon/M)\leqq C-\varepsilon$ for all n, another contradiction. This completes the proof of the theorem.

It is a consequence of the Fatou theorem and the fact that a monotone function has a finite derivative almost everywhere that a positive harmonic function $V(\zeta)$ in the upper half-plane takes finite boundary values at almost every point x on the real axis, that is to say, that

$$\lim_{r\to 0} V(x+re^{i\theta})$$

exists and equals $\mu'(x)$ uniformly for θ in closed subintervals of $(0,\pi)$. Every Pick function has the same property; to show this we consider an arbitrary Pick function and compose it with the linear fractional transformation $\psi(\zeta)$ which carries the upper half-plane into a circle of unit radius centered about the point $2+2i$. The composed function will have a positive real part as well as a positive imaginary part and so will have finite boundary values almost everywhere on the real axis. Composing this function with the inverse transformation ψ^{-1} will not affect the convergence of the function to a limit at almost all boundary points. The convergence which we describe here is a convergence in angle: $\varphi(x+re^{i\theta})$ converges, for almost all x, uniformly for θ in closed subintervals of $(0,\pi)$ to an appropriate limit. The following theorem shows that this is the usual state of affairs for Pick functions.

Theorem IV. *Let $\varphi(\zeta)$ be a Pick function which converges to a finite limit along the imaginary axis; the function then converges to the same limit in angle.*

Proof. We consider the family of Pick functions indexed by small positive values of r

$$\varphi_r(\zeta)=\varphi(r\zeta)$$

and note that this family is uniformly bounded at $\zeta=i$. From the theorems of Chapter II we can therefore extract a convergent sequence of functions and this sequence converges uniformly on compact subsets of the half-plane to a function in the Pick class which is constant on the imaginary axis. Evidently only one constant can occur, viz. the limit which $\varphi(\zeta)$ takes as ζ approaches the origin along the imaginary axis. Thus there can be only one limiting function, and $\varphi(re^{i\theta})$ converges to the same limit.

Chapter V. The Spectral Theorem

The study of Pick functions has long been associated with certain problems in analysis, in particular the Moment Problem and the theory of the spectral representation of self-adjoint operators in Hilbert space. The traditional proof of the spectral theorem deduces it from the integral representation of Pick functions given in Chapter II, and this argument has the advantage that the self-adjoint operator under study need not be supposed continuous. This chapter is devoted to that proof.

Let $\mathscr{H}$ be a separable Hilbert space and $\mathscr{D}$ a dense linear subspace of $\mathscr{H}$. We suppose that H is a linear transformation defined on the domain $\mathscr{D}$ and taking its values in $\mathscr{H}$. The transformation is called self-adjoint if the following two properties are satisfied:

(i) for every pair of vectors u and v in $\mathscr{D}$ the equation

$$(H u, v)=(u, H v)$$

is valid.

(ii) If f and g are two vectors in $\mathscr{H}$ such that

$$(H u, f)=(u, g)$$

for every u in $\mathscr{D}$, then f is also in $\mathscr{D}$ and $H f=g$.

Evidently $(H u, u)$ is real for every u in $\mathscr{D}$.

Let $\zeta=\xi+i\eta$ be a point in the upper half-plane and u a vector of norm 1. Now

$$\begin{aligned}\|H u-\zeta u\|^2&=\|H u-\zeta u\|^2+(H u, u)^2-(H u, u)^2\\&=(H u, H u)+|\zeta|^2-2\operatorname{Re}[\bar{\zeta}(H u, u)]+(H u, u)^2-(H u, u)^2\\&=(H u, H u)-(H u, u)^2+|\zeta-(H u, u)|^2 .\end{aligned}$$

From the Schwartz inequality we have $(H u, u)^2 \leqq (H u, H u)(u, u)=(H u, H u)$ and therefore

$$\|H u-\zeta u\|^2 \geqq |\zeta-(H u, u)|^2$$

and since $(H u, u)$ is real this leads to

$$\|H u-\zeta u\| \geqq \eta\|u\|$$

an inequality evidently valid for all u in $\mathscr{D}$.

Let I be the identity operator on Hilbert space. It is easy to see that the operator $H-\zeta I$ taken on the domain $\mathscr{D}$, has no null space. It is important to notice that the range of this operator is dense in $\mathscr{H}$, since if f is a vector orthogonal to that range, then

$$(Hu, f) = (u, \bar{\zeta} f)$$

for all u in $\mathscr{D}$. From (ii) above it then follows that f is in $\mathscr{D}$ and that $Hf = \bar{\zeta} f$, whence (Hf, f) is not real, a contradiction. It is now clear that $H-\zeta I$ has an inverse operator with a dense domain; we write it R_ζ and note that $\|R_\zeta f\| \leqq \|f\|/\eta$ for every f in its domain.

We next check that the range of $H-\zeta I$ is closed, i.e., that it is the whole space $\mathscr{H}$. Suppose f_n is a sequence of vectors in that range converging to a vector f_0. We write $f_n = (H-\zeta I)u_n$ and note that the u_n form a Cauchy sequence, since

$$\|u_n - u_m\| = \|R_\zeta(f_n - f_m)\| \leqq \|f_n - f_m\|/\eta\,.$$

So that sequence converges to an element u_0 and for every v in $\mathscr{D}$ the numbers

$$(v, f_n) = (v, (H-\zeta I)u_n) = ((H-\bar{\zeta} I)v, u_n)$$

converge to

$$(v, f_0) = ((H-\bar{\zeta} I)v, u_0)$$

and it follows again from (ii) that u_0 is in $\mathscr{D}$ and $(H-\zeta I)u_0 = f_0$. Hence R_ζ is a continuous linear transformation of the whole space $\mathscr{H}$ of bound $1/\eta$.

The transformation R_ζ is called the resolvent of H and is defined similarly for ζ in the lower half-plane. Its adjoint R_ζ^* is easily computed. For every u and v in $\mathscr{D}$ we have

$$(u, v) = ((H-\zeta I)R_\zeta u, v) = (R_\zeta u, (H-\bar{\zeta} I)v) = (u, R_\zeta^*(H-\bar{\zeta} I)v)$$

and therefore $R_\zeta^*(H-\bar{\zeta} I) = I$, whence $R_\zeta^* = (H-\bar{\zeta} I)^{-1} = R_{\bar{\zeta}}$. The resolvent also satisfies an important identity:

$$R_z - R_\zeta = (z-\zeta) R_z R_\zeta\,.$$

To establish this we apply $(H-zI)$ to the left side, obtaining

$$(H-zI)R_z - (H-zI)R_\zeta = I - (H-\zeta I + \zeta I - zI)R_\zeta = (z-\zeta)R_\zeta\,.$$

Multiplying now on the left by R_z we obtain the identity. From this it follows that R_z depends continuously on z, for if z and ζ vary over some compact set having a positive distance d from the real axis then

$$\|R_z - R_\zeta\| \leqq |z-\zeta|/d^2\,.$$

In particular, then, every function of the form $(R_\zeta u, v)$ is continuous in the open upper and lower half-planes. It is important to notice that such functions are even analytic since we have

$$\frac{(R_z u, v)-(R_\zeta u, v)}{z-\zeta}=(R_z R_\zeta u, v)$$

and if ζ is held fixed and z tends to ζ the right side tends continuously to $(R_\zeta^2 u, v)$ idependently of the way z approaches ζ. Thus the function is differentiable with respect to the complex variable, i.e., is an analytic function.

We have already had occasion to note that $(R_\zeta u, u)$ has a positive imaginary part in the upper half-plane, since, putting $R_\zeta u=f$ we have $Hf-\zeta f=u$ and so

$$(R_\zeta u, u)=(f, u)=(f, Hf)-\bar{\zeta}(f, f).$$

It follows that $(R_\zeta u, u)$ is a Pick function satisfying the inequality

$$|(R_\zeta u, u)| \leqq \|u\|^2/\eta .$$

Let us take u so that $\|u\|=1$ and consider $V(\zeta)$, the imaginary part of $(R_\zeta u, u)$. Now $V(i\eta) \leqq 1/\eta$ and so

$$\eta V(i\eta)=\alpha\eta^2+\int \frac{\eta^2}{\lambda^2+\eta^2} d\mu(\lambda)$$

is uniformly bounded by 1 for positive η. Evidently $\alpha=0$ and as η becomes very large the integrand above converges increasingly to $+1$. We infer that the measure μ has finite total mass.

A little more effort is required to show that the measure has total mass 1. We must write

$$u=(H-\zeta I) R_\zeta u=H R_\zeta u-\zeta R_\zeta u$$

to get

$$-\zeta R_\zeta u=u-H R_\zeta u$$

and therefore for every v in $\mathscr{H}$

$$(-\zeta R_\zeta u, v)=(u, v)-(H R_\zeta u, v).$$

If we take v in $\mathscr{D}$ this can be written

$$(-\zeta R_\zeta u, v)=(u, v)-(R_\zeta u, H v)$$

and as ζ approaches infinity along the imaginary axis the second term converges to 0. Since $\mathscr{D}$ is dense in $\mathscr{H}$ this implies that $-\zeta R_\zeta u$ converges weakly to u, from which it follows that $\eta V(i\eta)$ converges to $(u, u)=1$. However, it follows from the Fatou theorem in integration theory that the limit in question is also the total mass of the measure μ.

We next turn to an important example of a self-adjoint operator and its resolvent. Let μ be a positive Borel measure on the real axis and $L^2(\mu)$ the corresponding L^2 space. Let $\mathcal{D}$ be the subspace of $L^2(\mu)$ consisting of those functions $f(\lambda)$ for which $\lambda f(\lambda)$ is also in $L^2(\mu)$. When the measure is supported by a bounded subset of the axis it is obvious that $\mathcal{D}$ will coincide with $L^2(\mu)$, but otherwise this is not the case. It is evident that $\mathcal{D}$ contains the characteristic function of every finite interval and therefore that $\mathcal{D}$ is dense in $L^2(\mu)$. We define an operator H with domain $\mathcal{D}$ by setting

$$(Hf)(\lambda)=\lambda f(\lambda)$$

for every f in $\mathcal{D}$. It is quite easy to verify that H is self-adjoint, since if u and v are both in $\mathcal{D}$ then

$$(Hu,v)=\int \lambda u(\lambda)\overline{v(\lambda)}\,d\mu(\lambda)=\int u(\lambda)\lambda\overline{v(\lambda)}\,d\mu(\lambda)=(u,Hv);$$

while if

$$(Hu,f)=\int \lambda u(\lambda)\overline{f(\lambda)}\,d\mu(\lambda)=(u,g)=\int u(\lambda)\overline{g(\lambda)}\,d\mu(\lambda)$$

for every u in $\mathcal{D}$ then we must have $g(\lambda)=\lambda f(\lambda)$ almost everywhere, whence $\lambda f(\lambda)$ is in $L^2(\mu)$, that is, f is in $\mathcal{D}$ and $Hf=g$.

For every fixed ζ in the upper or lower half-plane define the operator R_ζ by the equation

$$(R_\zeta u)(\lambda)=\frac{1}{\lambda-\zeta}u(\lambda).$$

Since the function $1/(\lambda-\zeta)$ is bounded on the real axis R_ζ is a bounded linear transformation of $L^2(\mu)$ into itself, and it is easy to see that $\|R_\zeta\|=1/\eta$ where $\eta=\operatorname{Im}[\zeta]$. Moreover, the family of operators R_ζ satisfies the resolvent equation

$$R_z-R_\zeta=(z-\zeta)R_zR_\zeta$$

since the corresponding functions do. If u is in the domain of H and ζ a point in the upper half-plane, then

$$((H-\zeta I)u)(\lambda)=(\lambda-\zeta)u(\lambda)$$

and

$$(R_\zeta(H-\zeta I)u)(\lambda)=u(\lambda).$$

Thus R_ζ, as we have defined it, is in fact the resolvent of the self-adjoint H.

The Borel measure μ in this example is a positive measure defined on the class of all Borel subsets of the real axis. Another positive Borel measure v is said to be absolutely continuous relative to μ if and only if the family of Borel sets of v-measure zero contains the Borel sets of μ-measure zero. When this is the case there exists a nonnegative Borel measurable function $(dv/d\mu)(\lambda)$ on the axis so that for every Borel set E

$$\nu(E) = \int_E \frac{d\nu}{d\mu}(\lambda)\, d\mu(\lambda)$$

where of course the nonnegative integral may be infinite. The function $d\nu/d\mu$ is the Radon-Nikodym derivative of ν with respect to μ. Two positive Borel measures are equivalent if each is absolutely continuous relative to the other; in this case we will have

$$\frac{d\mu}{d\nu}(\lambda)\frac{d\nu}{d\mu}(\lambda) = 1$$

almost everywhere relative to either measure.

When the measures μ and ν are equivalent there exists an important linear transformation between the corresponding L^2 spaces. To the function $f(\lambda)$ in $L^2(\mu)$ we make correspond the function

$$(Uf)(\lambda) = f(\lambda)\sqrt{\frac{d\mu}{d\nu}(\lambda)}$$

which is evidently in $L^2(\nu)$ since

$$\int |(Uf)(\lambda)|^2\, d\nu(\lambda) = \int |f(\lambda)|^2\, d\mu(\lambda)\,.$$

Thus we have $\|Uf\| = \|f\|$ and the transformation U is an isometry. The mapping is onto since the inverse transformation is evidently given by

$$(U^{-1}g)(\lambda) = g(\lambda)\sqrt{\frac{d\nu}{d\mu}(\lambda)}$$

and is an isometry from $L^2(\nu)$ to $L^2(\mu)$.

Let $\mathscr{D}_1$ be the subspace of $L^2(\mu)$ consisting of functions $f(\lambda)$ for which $\lambda f(\lambda)$ is also in $L^2(\mu)$, and let H_1 be the self-adjoint operator with domain $\mathscr{D}_1$ corresponding to multiplication by λ. In a similar way we define a self-adjoint H_2 with domain $\mathscr{D}_2$ in $L^2(\nu)$. It is now clear that $\mathscr{D}_2 = U\mathscr{D}_1$ and $\mathscr{D}_1 = U^{-1}\mathscr{D}_2$, and it is also easy to see that $H_2 = UH_1U^{-1}$ and $H_1 = U^{-1}H_2U$. Thus H_1 and H_2 are unitarily equivalent.

It is our object here to show that an arbitrary self-adjoint operator on a separable Hilbert space is unitarily equivalent to a direct sum of operators H_i corresponding to multiplication by λ on a direct sum of Hilbert spaces of the form $L^2(\mu_i)$ for an appropriate sequence of positive Borel measures μ_i and thus that the example which we have studied is in some sense the most general self-adjoint operator. For this purpose in the sequel we shall suppose that there is given once and for all a self-adjoint H on a separable Hilbert space $\mathscr{H}$ with domain $\mathscr{D}$ and resolvent R_ζ.

Let f be an arbitrary element of $\mathscr{H}$ of norm 1 and let $\mathscr{M}_0(f)$ be the subspace consisting of all finite linear combinations of the vector $R_z f$ for non-real z. Every element of $\mathscr{M}_0(f)$ is therefore a finite sum of the form

$$h=\sum a_j R_{z_j} f.$$

By $\mathscr{M}(f)$ we denote the closure of $\mathscr{M}_0(f)$. It is easy to see that the operators R_z map $\mathscr{M}(f)$ into itself. Moreover, if g is orthogonal to $\mathscr{M}(f)$ we have $(R_z h, g)=0$ for all h in $\mathscr{M}(f)$, and therefore $(h, R_{\bar{z}} g)=0$. Thus $R_{\bar{z}} g$ is also orthogonal to $\mathscr{M}(f)$. It follows that the orthogonal decomposition

$$\mathscr{H}=\mathscr{M}(f)\oplus\mathscr{M}(f)^{\perp}$$

reduces R_z for every non-real value of z.

Let $\mathscr{L}$ be the linear space of all formal finite sums

$$h=\sum a_i R_{z_i}$$

and T_1 the linear mapping from $\mathscr{L}$ to $\mathscr{M}(f)$ which carries h into

$$T_1 h=\sum a_i R_{z_i} f .$$

Let μ be the measure occuring in the canonical representation of the Pick function

$$(R_\zeta f, f)=\int \frac{d\mu(\lambda)}{\lambda-\zeta}$$

and T_2 the linear mapping from $\mathscr{L}$ to $L^2(\mu)$ which carries h into

$$(T_2 h)(\lambda)=\sum a_i \frac{1}{\lambda-z_i}.$$

It is important to notice that $\|T_1 h\|=\|T_2 h\|$ for every h in $\mathscr{L}$ as the following calculation shows. We have

$$\|T_1 h\|^2=\sum\sum a_j \bar{a}_k (R_{z_j} f, R_{z_k} f)=\sum\sum a_j \bar{a}_k (R_{\bar{z}_k} R_{z_j} f, f).$$

Moreover, from the resolvent equation

$$R_{\bar{z}_k} R_{z_j}=\frac{1}{\bar{z}_k-z_j}(R_{\bar{z}_k}-R_{z_j})$$

and therefore

$$\begin{aligned}(R_{\bar{z}_k} R_{z_j} f, f)&=\frac{1}{\bar{z}_k-z_j}\int\left[\frac{1}{\lambda-\bar{z}_k}-\frac{1}{\lambda-z_j}\right] d\mu(\lambda)\\ &=\int \frac{1}{\lambda-z_j}\,\overline{\frac{1}{\lambda-z_k}}\, d\mu(\lambda).\end{aligned}$$

Accordingly

$$\|T_1 h\|^2 = \sum\sum a_j \bar{a}_k \int \frac{1}{\lambda - z_j} \overline{\frac{1}{\lambda - z_k}} \, d\mu(\lambda)$$
$$= \int |(T_2 h)(\lambda)|^2 \, d\mu(\lambda) = \|T_2 h\|^2 .$$

We also note that the range of T_1 is the dense subspace $\mathcal{M}_0(f)$ of $\mathcal{M}(f)$, and now prove that the range of T_2 is dense in $L^2(\mu)$. For this purpose we suppose that $g(\lambda)$ is a function in $L^2(\mu)$ orthogonal to all functions of the form $1/(\lambda - z)$ for non-real z. Evidently the function

$$N(z) = \int \frac{g(\lambda)\, d\mu(\lambda)}{\lambda - z}$$

vanishes for all non-real z. We can write $g(\lambda) = u(\lambda) + i v(\lambda)$ where u and v are real functions to obtain

$$\int \frac{u(\lambda) + i v(\lambda)}{\lambda - z} \, d\mu(\lambda) = 0$$

and passing to $\bar{z}$ we have also

$$\int \frac{u(\lambda) + i v(\lambda)}{\lambda - \bar{z}} \, d\mu(\lambda) = \int \overline{\frac{u(\lambda) - i v(\lambda)}{\lambda - z}} \, d\mu(\lambda) = 0$$

and so we may infer that

$$\int \frac{u(\lambda)}{\lambda - z} \, d\mu(\lambda) = 0$$

for the real, μ-integrable function $u(\lambda)$. If we now write $u(\lambda)$ as the difference of two positive functions $u_+(\lambda)$ and $u_-(\lambda)$ we finally obtain an equality between Pick functions:

$$\int \frac{u_+(\lambda)\, d\mu(\lambda)}{\lambda - z} = \int \frac{u_-(\lambda)\, d\mu(\lambda)}{\lambda - z}$$

and from the uniqueness of the measure in the canonical representation of a Pick function we infer that $u_+(\lambda) = u_-(\lambda)$ almost everywhere relative to the measure μ. Hence $u(\lambda) = 0$ almost everywhere and a similar argument shows that $v(\lambda) = 0$ almost everywhere. This makes $g(\lambda)$ the 0 element of $L^2(\mu)$ as desired.

From the equation $\|T_1 h\| = \|T_2 h\|$ it follows that the transformations T_1 and T_2 have a common null-space $\mathcal{N}$ in $\mathcal{L}$. Let S_1 be the quotient mapping, carrying $\mathcal{L}/\mathcal{N}$ into a dense subspace of $\mathcal{M}(f)$, and let S_2 be the corresponding quotient mapping into a dense subspace of $L^2(\mu)$. It is now plain that $S_2 S_1^{-1}$ determines a unitary mapping V from $\mathcal{M}(f)$ to $L^2(\mu)$ and it is also clear that $(V R_z f)(\lambda) = 1/(\lambda - z)$. We have already seen that $-z R_z f$ converges weakly to f in $\mathcal{H}$ and therefore in $\mathcal{M}(f)$, and from this it follows that $(Vf)(\lambda)=+1$, i.e., the image of f is the constant function in the L^2 space. The elements of $\mathcal{M}(f)$ which are in the domain of H are exactly the elements in the range of R_z for some fixed z. Under the mapping V these go into exactly those functions of $L^2(\mu)$ of the form $h(\lambda)/(\lambda-z)$ where $h(\lambda)$ is in $L^2(\mu)$, and therefore are exactly those functions $f(\lambda)$ with the property that $\lambda f(\lambda)$ is in $L^2(\mu)$. Clearly, the self-adjoint operator H, when restricted to that part of its domain in $\mathcal{M}(f)$, is unitarily equivalent by the mapping V to the operation of multiplication by λ in $L^2(\mu)$.

We can now obtain a preliminary description of the operator H by choosing an element f_1 in $\mathcal{H}$ with $\|f_1\|=1$ and constructing the corresponding $\mathcal{M}(f_1)$ and $L^2(\mu_1)$ so that H, when considered on $\mathcal{M}(f_1)$, is unitarily equivalent to multiplication by λ on $L^2(\mu_1)$. We then pass to the orthogonal complement of $\mathcal{M}(f_1)$ and choose there a normalized element f_2. We form $\mathcal{M}(f_2)$ and the corresponding $L^2(\mu_2)$. This process can be continued, and there will therefore exist orthogonal decompositions of $\mathcal{H}$ of the form

$$\mathcal{H}=\mathcal{M}(f_1)\oplus\mathcal{M}(f_2)\oplus\mathcal{M}(f_3)\oplus\cdots\oplus\mathcal{H}^*$$

where $\mathcal{H}^*$ is also a reducing subspace for H. We appeal to Zorn's Lemma to obtain a maximal decomposition, i.e., a representation of $\mathcal{H}$ as a finite or countable direct sum of spaces of the $\mathcal{M}(f_i)$. Evidently the operator H is unitarily equivalent to the operation of multiplication by λ on the direct sum of the corresponding $L^2(\mu_i)$.

We have therefore obtained the following result: every self-adjoint H on a separable Hilbert space is unitarily equivalent to the operation of multiplication by λ on a direct sum of L^2 spaces on the real axis. We must make this representation a canonical one. For this purpose we should first notice certain facts which permit us to simplify the representation. Suppose the space has been decomposed as a direct sum of subspaces $\mathcal{M}(f_i)$. The normalized elements f_i form an orthonormal set. Choose coefficients a_i with $a_i>0$ such that $\sum a_i^2=1$ and define an element $f_0=\sum a_i f_i$. Now

$$(R_z f_0, f_0)=\sum a_i^2(R_z f_i, f_i)=\int\frac{1}{\lambda-z}\,d\left(\sum a_i^2\mu_i\right)(\lambda)$$

and so

$$(R_z f_0, f_0)=\int\frac{1}{\lambda-z}\,d\mu_0(\lambda)$$

where $\mu_0=\sum a_i^2\mu_i$. It is now obvious that every measure μ_i is absolutely continuous relative to μ_0. From this it easily follows that for every element u in the Hilbert space the measure occuring in the canonical representation of the Pick function $(R_\zeta u, u)$ is absolutely continuous relative to μ_0. Indeed, we may regard the Hilbert space as a direct sum of spaces $L^2(\mu_i)$ and take u_i as the projection of u on the i-th space. Then

$$(R_\zeta u, u) = \sum_{i=1}^{\infty} (R_\zeta u_i, u_i) = \sum_{i=1}^{\infty} \int \frac{1}{\lambda - z} |u_i(\lambda)|^2 \, d\mu_i(\lambda)$$

and every measure appearing in the sum is absolutely continuous relative to μ_0.

We should emphasize that μ_0 is uniquely determined up to an equivalent measure; for every f in $\mathscr{H}$ the function $(R_\zeta f, f)$ corresponds to a measure μ which is absolutely continuous relative to μ_0 and μ_0 is in fact obtained from a Pick function of this form, namely $(R_\zeta f_0, f_0)$. It is obvious that any other measure having these properties is equivalent to μ_0.

Since the measures μ_i are absolutely continuous relative to μ_0 we may use the Radon-Nikodym derivative and write

$$\mu_i(E) = \int_E \frac{d\mu_i}{d\mu_0}(\lambda) \, d\mu_0(\lambda)$$

for every Borel set E. Let us put $\chi_i(\lambda)$ as the characteristic function of the set E_i where $(d\mu_i/d\mu_0)(\lambda) > 0$ and define a new measure μ_i' by putting

$$\mu_i'(E) = \int_E \chi_i(\lambda) \, d\mu_0(\lambda)$$

for every Borel set E. Since the measures μ_i and μ_i' are obviously equivalent measures, we are at liberty to represent H in terms of measures of the form μ_i' which are restrictions of μ_0 to certain Borel subsets E_i of the real axis. So far, the representation so obtained is in no sense unique, since there may be a variety of decompositions of the Hilbert space into spaces of the form $\mathscr{M}(f_i)$. However, we can now find an entity which is uniquely determined: the multiplicity function. We define this function as follows:

$$M(\lambda) = \sum \chi_i(\lambda)$$

where the $\chi_i(\lambda)$ are the Radon-Nikodym derivatives, which, as we have seen, may be taken to be characteristic functions of certain Borel sets E_i. We make the convention that $M(\lambda)$ may take the value $+\infty$ on a set of positive μ_0-measure. In any event, the values of $M(\lambda)$ are always nonnegative integers and $+\infty$.

Lemma 1. *The multiplicity function $M(\lambda)$ is uniquely determined up to a set of μ_0-measure* 0.

Proof. If the lemma is false there exists an operator H which has two distinct representations, necessarily unitarily equivalent, involving sequences of characteristic functions $\chi'_i(\lambda)$ and $\chi''_i(\lambda)$ such that the sums

$$M'(\lambda)=\sum \chi'_i(\lambda) \quad \text{and} \quad M''(\lambda)=\sum \chi''_i(\lambda)$$

differ on a bounded Borel set E with $\mu_0(E)>0$. The functions $M'(\lambda)$ and $M''(\lambda)$ cannot both be infinite on E, and each takes at most countably many values. It is then easy to see that we may assume that one of the functions, say $M'(\lambda)$, is exactly equal to some positive integer m on E, while $M''(\lambda)$ is at least $m+1$ on that set. There is also no loss of generality in our supposing that the sequences of measures have been so enumerated that $\mu'_i(E)>0$ for $i\leqq m$, whence $\mu'_i(E)=0$ for $i>m$, and that $\mu''_i(E)>0$ for $i\leqq m+1$. Let h''_j be the function in $L^2(\mu''_i)$ defined by the conditions

$$\begin{aligned} h''_i(\lambda) &= 1 \quad \text{for } \lambda \text{ in } E \\ &= 0 \quad \text{otherwise}. \end{aligned}$$

We consider only values of i in the interval $1\leqq i\leqq m+1$. Let W be the unitary mapping which carries the representation of H relative to the spaces $L^2(\mu''_i)$ into the representation corresponding to the spaces $L^2(\mu'_i)$. We put

$$g'_i = W h''_i = \sum_{k=1}^{\infty} g'_{ik},$$

where g'_{ik} is the projection of g'_i on $L^2(\mu'_k)$, and note that

$$(R_z h''_i, h''_i) = \int_E \frac{1}{\lambda - z}\, d\mu_0(\lambda) = (R_z g'_i, g'_i) = \sum_{k=1}^{\infty} \int \frac{1}{\lambda - z} |g'_{ik}(\lambda)|^2\, d\mu_0(\lambda).$$

From this we infer that the functions $g'_{ik}(\lambda)$ vanish outside E, since the measure occuring in the canonical representation of a Pick function is unique, while for almost all λ in that set we have

$$\sum_{k=1}^{\infty} |g'_{ik}(\lambda)|^2 = 1.$$

It is more important to notice that there are at most m terms in this sum, since $\mu'_i(E)=0$ for $i>m$.

Now let i be different from j; for all non-real z we have $(R_z h''_i, h''_j)=0$ and therefore $(R_z g'_i, g'_j)=0$, and this may be written explicitly

$$\sum_{k=1}^{m} (R_z g'_{ik}, g'_{jk}) = \sum_{k=1}^{m} \int \frac{g'_{ik}(\lambda)\overline{g'_{jk}(\lambda)}}{\lambda - z}\, d\mu_0(\lambda) = 0.$$

We next define a real measure ν_{ij} on the axis by putting

$$d\nu_{ij}(\lambda) = \sum_{k=1}^{m} g'_{ik}(\lambda)\,\overline{g'_{jk}(\lambda)}\,d\mu_0(\lambda);$$

this is of finite total mass since the functions $g_{ik}(\lambda)$ are integrable square relative to μ_0. Our equation reduces to

$$\int \frac{d\nu_{ij}(\lambda)}{\lambda - z} = 0$$

valid for all non-real z, and so, in view of an argument given earlier in this chapter, we infer that the measure ν_{ij} vanishes identically. From this it follows that for i different from j

$$\sum_{k=1}^{m} g'_{ik}(\lambda)\,\overline{g'_{jk}(\lambda)} = 0$$

almost everywhere relative to the measure μ_0, while we have already seen that

$$\sum_{k=1}^{m} |g'_{ik}(\lambda)|^2 = 1$$

almost everywhere relative to that measure. Thus for at least one value of λ these equations are satisfied, and they determine an orthonormal system of $m+1$ elements in an m-dimensional Hilbert space. This contradiction completes the proof.

Lemma 2. *If* $M(\lambda) = +\infty$ *almost everywhere relative to the measure* μ_0 *then* H *admits a representation as multiplication by* λ *on a direct sum of countably many copies of* $L^2(\mu_0)$.

Proof. We have

$$M(\lambda) = \sum \chi_i(\lambda) = +\infty$$

for some representation of H on a direct sum of spaces $L^2(\mu_i)$. For each index i we decompose the characteristic function as follows:

$$\chi_i(\lambda) = \chi'_i(\lambda) + \chi''_i(\lambda),$$

where

$$\chi'_i(\lambda) = \chi_i(\lambda) \prod_{j<i} [1 - \chi_j(\lambda)].$$

This gives rise to a corresponding decomposition of the measure:

$$\mu_i = \mu'_i + \mu''_i$$

where

$$\mu'_i(E) = \int_E \chi'_i(\lambda)\,d\mu_0(\lambda)$$

for every Borel set E. Accordingly, the L^2 space has the orthogonal decomposition

$$L^2(\mu_i) = L^2(\mu_i') \oplus L^2(\mu_i'') .$$

It is clear that $\sum \chi_i'(\lambda) = +1$ almost everywhere for μ_0 and therefore that the direct sum of the spaces $L^2(\mu_i')$ is the same thing as $L^2(\mu_0)$. The Hilbert space is now a direct sum of $L^2(\mu_0)$ and the spaces $L^2(\mu_i'')$, and we should note that the first factor contains $L^2(\mu_1)$. We can repeat the process with the direct sum of the $L^2(\mu_i)$ to sort out another factor of the form $L^2(\mu_0)$, representing the space in the form

$$L^2(\mu_0) \oplus L^2(\mu_0) \oplus \mathscr{H}_2$$

where $\mathscr{H}_2$ is a direct sum of spaces $L^2(\mu_i''')$ and $L^2(\mu_2)$ is a subset of $L^2(\mu_0) \oplus L^2(\mu_0)$. By induction we have for each k a decomposition of the space into two factors, one involving a direct sum of k copies of $L^2(\mu_0)$ and containing the spaces $L^2(\mu_j)$ for $j \leqq k$, and the other a direct sum of L^2-spaces with an infinite multiplicity function. It follows from this that our space is in fact a countable sum of copies of $L^2(\mu_0)$.

We are now able to complete our description of the operator H. We may suppose f_0 chosen in $\mathscr{H}$ with $\|f_0\| = 1$ and $\mathscr{M}(f_0)$ corresponding to a positive measure μ_0 of total mass 1. The multiplicity function $M(\lambda)$ is $\geqq 1$ almost everywhere relative to the measure μ_0. If we pass next to the orthogonal complement of $\mathscr{M}(f_0)$ and consider the operator family R_z on that space we may select a normalized f_1 in that space corresponding to a measure μ_1 which has two properties: (1) for any other vector g in that space, the corresponding measure is absolutely continuous relative to μ_1, and (2) μ_1 is the restriction of μ_0 to a Borel subset E_1 of the real axis. It is easy to identify E_1 if we notice that the the multiplicity function $M_1(\lambda)$ corresponding to the orthogonal complement of $\mathscr{M}(f_0)$ satisfies the equation

$$M(\lambda) = 1 + M_1(\lambda) \quad \text{almost everywhere } \mu_0 .$$

Thus E_1 is the set where $M(\lambda) \geqq 2$. Again, in the orthogonal complement of the direct sum $\mathscr{M}(f_0) \oplus \mathscr{M}(f_1)$ we can select a normalized element f_2 corresponding to a measure μ_2 which is the restriction of μ_0 to the set $M(\lambda) \geqq 3$. In this way we obtain a sequence of subspaces $\mathscr{M}(f_k)$ corresponding to a sequence of measures μ_k with μ_k the restriction of μ_0 to the set $M(\lambda) \geqq k+1$. Either the direct sum of the spaces $\mathscr{M}(f_k)$ exhausts the Hilbert space $\mathscr{H}$ or there remains a factor space $\mathscr{H}^*$ corresponding to a measure μ^* and a multiplicity function which is identically infinite. From Lemma 2 it follows that $\mathscr{H}^*$ is a countable direct sum of spaces of the form $\mathscr{M}(g_k)$ all corresponding to the same measure, namely μ^*. We have finally a form of the spectral theorem.

Theorem I. *A self-adjoint operator H acting on a separable Hilbert space is completely characterized up to unitary equivalence by the equivalence class of μ_0, a positive Borel measure on the real axis, and the corresponding multiplicity function $M(\lambda)$. The operator is unitarily equivalent to the operation of multiplication by λ on a direct sum of spaces $L^2(\mu_k)$ where μ_k is the restriction of μ_0 to a Borel set E_k and $E_{k+1} \subseteq E_k$ for all k.*

Chapter VI. One-Dimensional Perturbations

Let A be a symmetric transformation of an n-dimensional Hilbert space into itself and P the projection of that space onto the one-dimensional subspace determined by the normalized vector f. Our principal interest in this chapter is to investigate the spectrum of the operator

$$B = A + cP$$

where $c > 0$ is given.

Suppose e_k is an eigenvector of A which is orthogonal to f; evidently $Pe_k = 0$ and so $Ae_k = Be_k = \xi_k e_k$. Thus e_k is also an eigenvector for B with the same eigenvalue.

Let the spectrum of A be written

$$\xi_1 \leqq \xi_2 \leqq \xi_3 \leqq \cdots \leqq \xi_n$$

where each eigenvalue is taken as often as its multiplicity requires. Let $\{e_i\}$ be the system of corresponding eigenvectors. If an eigenvalue ξ_k occurs with multiplicity greater than 1, we may select the corresponding eigenvectors in such a way that all, or at least all but one of them, are orthogonal to f.

Let R_ζ be the resolvent operator for A, i.e., $R_\zeta = (A - \zeta I)^{-1}$ and R'_ζ the resolvent for $B: R'_\zeta = (B - \zeta I)^{-1}$. If $R'_\zeta u = v$ then

$$(B - \zeta I)^{-1} u = v.$$

$$(B - \zeta I) v = u.$$

$$Av + cPv - \zeta v = u.$$

$$(A - \zeta I) v = u - c(v, f) f.$$

Now apply the resolvent R_ζ to obtain

$$v = R_\zeta u - c(v, f) R_\zeta f.$$

We use this relation to determine (v, f) as follows:

$$(v, f) = (R_\zeta u, f) - c(v, f)(R_\zeta f, f)$$

whence

$$(v, f) = \frac{(R_\zeta u, f)}{1 + c(R_\zeta f, f)}.$$

Accordingly,
$$R'_\zeta u = v = R_\zeta u - \frac{(R_\zeta u, f)}{\frac{1}{c} + (R_\zeta f, f)} R_\zeta f .$$

This equation gives the relationship between R'_ζ and R_ζ. Now the eigenvalues of B are clearly the singularities of R'_ζ, and these are either poles inherited from R_ζ, or poles arising from zeros in the Pick function

$$\varphi(\zeta) = \frac{1}{c} + (R_\zeta f, f) = \frac{1}{c} + \sum_{i=1}^{n} \frac{|(f, e_i)|^2}{\xi_i - \zeta} .$$

It is evident that the poles of the Pick function $\varphi(\zeta)$ are a subset of the eigenvalues of A, while its zeros are a subset of the eigenvalues of B. It is also clear that there is a zero between each pole, as well as a zero to the right of the largest pole, the function being positive and regular at infinity. We write these zeros and poles in order:

$$\xi'_1 < \eta'_1 < \xi'_2 < \eta'_2 < \cdots < \xi'_m < \eta'_m$$

where the η' denote the zeros. Now adjoin to this list those common eigenvalues of A and B arising from eigenvectors e_k orthogonal to f. In each case we interpolate above a pair $\xi = \eta$. Thus, the spectra of A and B are related in the following way:

$$\xi_1 \leqq \eta_1 \leqq \xi_2 \leqq \eta_2 \leqq \cdots \leqq \xi_n \leqq \eta_n .$$

We should also note that

$$\operatorname{Trace}(B - A) = \operatorname{Trace}(cP) = c = \operatorname{Trace} B - \operatorname{Trace} A = \sum_{k=1}^{n} (\eta_k - \xi_k) .$$

For a future application it is convenient now to note the following fact. Let points ξ_i and η_j be related so that

$$\xi_1 < \eta_1 < \xi_2 < \eta_2 < \cdots < \xi_n < \eta_n$$

where the ξ_i are the eigenvalues of a given symmetric matrix A. Then there exists $c > 0$ and a one-dimensional projection P such that $B = A + cP$ has exactly the numbers η_j for its spectrum. To show this we consider the function

$$\varphi(\zeta) = \prod_{k=1}^{n} \frac{\zeta - \eta_k}{\zeta - \xi_k}$$

a product of monotone increasing functions. Since $\varphi(\zeta)$ is evidently rational, of degree n, with n simple poles, it is a Pick function, since the residues at those poles are all negative. Accordingly, we can write it

$$\varphi(\zeta) = 1 + \sum_{j=1}^{n} \frac{m_j}{\xi_j - \zeta}$$

since the function is regular and equal to 1 at infinity. Put $c=\sum m_j$ as the total mass of the measure, writing

$$\varphi(\zeta)=1+c\sum_{j=1}^{n}\frac{m'_j}{\xi_j-\zeta}$$

where $\sum m'_j=1$. If $\{e_k\}$ is the set of eigenvectors for A, put $f=\sum a_k e_k$ where the complex number a_k is so chosen that $|a_k|^2=m'_k$. Let P be the projection corresponding to that choice of f, of norm 1. The operator $B=A+cP$ then has the desired spectrum.

The construction which we have just completed is so important that it is worthwhile to consider another aspect of it. The function $\varphi(\zeta)$ was of course completely determined by the specification of its zeros and poles, since its value at infinity is always $+1$. Thus the numbers m_i are also determined. These numbers are the solutions to the system of linear equations

$$\sum_{i=1}^{n}\frac{m_i}{\xi_i-\eta_j}=-1$$

and the solution is unique. We infer that the determinant of this system of inhomogenous equations is not zero. It is therefore easy to see, in view of our hypothesis concerning the order of the zeros and poles of $\varphi(\zeta)$ that

$$\det\left|\frac{1}{\eta_i-\xi_j}\right|>0$$

since the non-zero determinant is surely a continuous function of the $2n$ distinct variables ξ_j and η_i, while if the η are very close to the corresponding ξ, say $\eta_i=\xi_i+\varepsilon$, then the diagonal elements of the determinant are $1/\varepsilon$ which is very large and positive. The other elements of the matrix being bounded, it is clear that the determinant itself is positive.

Another way to establish this important fact is to compute the determinant explicitly in the same way that one computes the Vandermonde determinant. We claim that

$$\prod_{i,j}(\eta_i-\xi_j)\det\left|\frac{1}{\eta_i-\xi_j}\right|=(-1)^{n(n-1)/2}\prod_{k<i}(\xi_i-\xi_k)(\eta_i-\eta_k).$$

Each side of this identity is a polynomial in the $2n$ variables ξ,η which is of degree $n-1$ in each variable. However, whenever $\xi_i=\xi_j$ for $i\neq j$ the determinant evidently vanishes, so $(\xi_i-\xi_j)$ divides the left side. The same argument shows that $(\eta_i-\eta_j)$ divides the left side. Thus the two polynomials are the same, except, perhaps, for a constant factor obtained by dividing the right side into the left. We must show that this factor is $+1$.

On the left, multiply the polynomial into the determinant, to obtain the determinant of a matrix having the polynomial $\prod_{l \neq j}(\eta_i - \xi_l)$ in the i-th row and j-th column. Now let the ξ's and η's be equal: $\xi_i = \eta_i$ for all i. The matrix in question becomes a diagonal matrix with determinant

$$\prod_{i=1}^{n} \prod_{l \neq i} (\eta_i - \xi_l).$$

Half of the terms in this product are negative, so it can be written

$$(-1)^{n(n-1)/2} \prod_{k<i} (\eta_i - \eta_k)^2$$

and this is the polynomial on the right. From this identity we can also see that the determinant $\det|1/(\eta_i - \xi_j)|$ is strictly positive when the ξ's and η's are the poles and zeros of a rational Pick function, since the polynomial $\prod_{i,j}(\eta_i - \xi_j)$ has exactly $n(n-1)/2$ negative factors.

Since every symmetric transformation is a linear combination of one dimensional projections associated with the eigenvectors, the arguments of this chapter show that the spectrum depends continuously on the transformation. Indeed if A and H are symmetric transformations having respectively the eigenvalues $\xi_1 \leqq \xi_2 \leqq \cdots \leqq \xi_n$ and $\lambda_1 \leqq \lambda_2 \leqq \cdots \leqq \lambda_n$ then the passage from A to $A+H$ may be regarded as the successive application of n one-dimensional perturbations associated with the operators $\lambda_i P_i$ where P_i is the projection on the i-th normalized eigenvector of H. At the k-th step of this process the j-th eigenvalue of the unperturbed matrix is shifted at most by a distance $|\lambda_k|$ = absolute value of the trace of the perturbation. Hence if the eigenvalues of $A+H$ are written $\xi'_1 \leqq \xi'_2 \leqq \cdots \leqq \xi'_n$ we have

$$|\xi_j - \xi'_j| \leqq \sum |\lambda_i| \leqq \sqrt{n \sum \lambda_i^2} = \sqrt{n} \|H\|$$

where $\|H\|$ is the Hilbert-Schmidt norm of H which we shall consider in Chapter VIII.

Chapter VII. Monotone Matrix Functions

Let H be a bounded, self adjoint transformation on a Hilbert space with spectrum in the open interval (a,b) of the real axis. Suppose that $f(x)$ is a real function defined on the interval (a,b). One then defines the operator $f(H)$ from the spectral decomposition of H as follows:

$$\text{if } H=\int\lambda\,dE_\lambda \quad \text{then } f(H)=\int f(\lambda)\,dE_\lambda$$

and this definition makes sense provided $f(\lambda)$ satisfies certain measurability conditions. However, in the special case when the Hilbert space is finite dimensional, no measurability conditions need be imposed, and the definition which we have given above can be put in a substantially simpler, but equivalent form. One then simply writes H as a diagonal matrix with eigenvalues λ_i displayed on the diagonal; $f(H)$ then appears as another diagonal matrix, with the entry $f(\lambda_i)$ in the i-th place. Thus, H and $f(H)$ have the same eigenvectors, and if λ is an eigenvalue of H, then $f(\lambda)$ is an eigenvalue of $f(H)$.

It is now clear that if $f(x)$ is defined, real and finite on the interval (a,b), then there corresponds an operator function $f(H)$ defined for all symmetric transformations of a given finite-dimensional Hilbert space having spectra in the interval (a,b). This class of operators is an open set in the finite-dimensional operator space in view of results of the previous chapter. We use the same letter, f, to denote both the operator function and the function of the real variable x in (a,b), since no confusion can arise.

Although the function we have just described is properly called an operator function, we make an imprecise use of the language and follow the established custom of calling the function a matrix function. This traditional terminology at least has the advantage of emphasizing the fact that the associated Hilbert spaces are finite-dimensional.

Let $f(x)$ be a matrix function, defined for symmetric matrices of order n with spectra in the corresponding interval (a,b). The domain of f is a class of operators admitting a natural order: $A\leqq B$ if and only if $B-A$ is a positive matrix. Equivalently, $A\leqq B$ if and only if for every vector u in the associated n-dimensional Hilbert space $(Au,u)\leqq(Bu,u)$.

In this chapter, and in many subsequent chapters, we will be interested in the class of monotone matrix functions of order n: these are the matrix functions that preserve the operator ordering:

$$A \leqq B \quad \text{implies} \quad f(A) \leqq f(B).$$

We denote the class of monotone matrix functions of order n associated with the interval (a,b) by $P_n(a,b)$. Certain properties of this class are almost immediate.

(1) $P_1(a,b)$ is simply the class of all real, monotone increasing functions on the open interval, since the only symmetric transformations of a one-dimensional Hilbert space into itself with spectra in the interval (a,b) are simply the numbers of that interval.

(2) $P_n(a,b)$ is a convex cone: if f and g belong to $P_n(a,b)$ and α and β are positive numbers, then $\alpha f + \beta g$ is also in $P_n(a,b)$.

(3) If $f(x)$ belongs to $P_n(a,b)$ and takes values in some interval (c,d) and if $g(x)$ belongs to $P_n(c,d)$, then the composition $(g \circ f)(x) = g(f(x))$ is also in $P_n(a,b)$. The easy proof goes as follows: if $A \leqq B$ and both operators have their spectra in (a,b) then $f(A) \leqq f(B)$ and both these operators have their spectra in (c,d), whence $g(f(A)) \leqq g(f(B))$, as desired.

(4) $P_n(a,b)$ is closed in the topology of pointwise convergence.

To show this, we suppose that $f_k(x)$ is a sequence in $P_n(a,b)$ converging pointwise to a function $f_0(x)$ and show that the limit is also in $P_n(a,b)$. Suppose that A and B are symmetric operators, with spectra in the interval (a,b) for which $A \leqq B$. Let the eigenvectors of A be e'_i with eigenvalues ξ_i corresponding. For any vector u in the space we may write

$$u = \sum_i (u, e'_i) e'_i \quad \text{and} \quad Au = \sum_i (u, e'_i) \xi_i e'_i$$

whence

$$(f_k(A)u, u) = \sum_i f_k(\xi_i) |(u, e'_i)|^2 .$$

If the eigenvectors of B are e''_j with eigenvalues η_j this quantity is not larger than

$$(f_k(B)u, u) = \sum f_k(\eta_j) |(u, e''_j)|^2$$

and as k converges to infinity we obtain

$$(f_0(A)u, u) = \sum_i f_0(\xi_i) |(u, e'_i)|^2 \leqq \sum_j f_0(\eta_j) |(u, e''_j)|^2 = (f_0(B)u, u)$$

as desired.

(5) The linear function $\alpha x + \beta$ belongs to $P_n(a,b)$ for any interval and any value of n, provided that $\alpha \geqq 0$.

(6) $P_{n+1}(a,b)$ is a subset of $P_n(a,b)$.

Before giving the formal proof, we remark that, as the dimension n increases, the family of inequalities that f in $P_n(a,b)$ must satisfy grows larger, and therefore, the class grows smaller.

Let f belong to $P_{n+1}(a,b)$ and let A and B be operators on an n-dimensional Hilbert space such that $A \leqq B$. We extend these operators to a space of dimension $n+1$ by adjoining a vector v, orthogonal to the n-dimensional space, and defining $Av = Bv = cv$ where c is chosen in the interval (a,b). For the extended operators we of course have $f(A) \leqq f(B)$; descending to the n-dimensional subspace we find the same condition true.

(7) Let $f(x)$ belong to $P_n(a,b)$; its regularization $f_\varepsilon(x)$ then belongs to $P_n(a+\varepsilon, b-\varepsilon)$.

The regularization is clearly a pointwise limit of combinations with positive coefficients of translates of the function: $f(x+t)$ where $|t| \leqq \varepsilon$. In view of properties 2, 3, and 5 above, those translates and their combinations are also in $P_n(a+\varepsilon, b-\varepsilon)$ and therefore the pointwise limit is also.

The important theorem which follows makes it clear why we have chosen the notation $P_n(a,b)$ for the class of monotone matrix functions of order n associated with the interval (a,b).

Theorem I. *Let $\varphi(\zeta)$ belong to $P(a,b)$; then this function belongs to $P_n(a,b)$ for every n.*

Proof. The function $\varphi(\zeta)$ in $P(a,b)$ is surely the limit of functions of the form

$$\varphi_N(\zeta) = \alpha\zeta + \beta + \int_{-N}^{+N} \frac{d\mu(\lambda)}{\lambda - \zeta}$$

where the measure μ puts no mass in the open interval (a,b). Because $P_n(a,b)$ is closed under pointwise convergence, it is enough to show that these functions are in $P_n(a,b)$. We already know that that class is a convex cone, and that linear functions of the form $\alpha\zeta + \beta$ belong to it; hence, only the integral need be considered in our proof. Now these integrals are limits of rational functions in the Pick class:

$$\varphi(\zeta) = \sum \frac{m_i}{\lambda_i - \zeta}$$

where either $\lambda_i < a$ or $\lambda_i > b$. It follows that it is enough to show that $(\lambda - \zeta)^{-1}$ is in $P_n(a,b)$, provided $\lambda < a$ or $\lambda > b$. This, in turn, will be a consequence of the following lemma, which we prove presently.

Lemma. *The function $-1/\zeta$ belongs to $P_n(0,\infty)$ and to $P_n(-\infty,0)$ for all n.*

Accordingly, if $\lambda < a$, the function $\zeta - \lambda$ belongs to $P_n(a,b)$ and is positive; we compose it with $-1/\zeta$ to obtain $(\lambda-\zeta)^{-1}$ in $P_n(a,b)$. Similarly, for $\lambda > b$, $\zeta-\lambda$ is negative on (a,b) and can be composed with $-1/\zeta$ to put $(\lambda-\zeta)^{-1}$ in $P_n(a,b)$.

Proof of Lemma. If the operator A on a finite dimensional Hilbert space has its spectrum in the interval $\lambda > 0$ it does not have 0 has an eigenvalue and therefore has an inverse A^{-1}. It also has a positive square root: $S>0$ with $S^2 = A$ where S is defined by means of the smooth function $\sqrt{x}$ on $x>0$. Clearly $(S^{-1})^2 = A^{-1}$. We may write, using the Schwartz inequality

$$|(x,y)|^2 = |(S^{-1}x, Sy)|^2 \leqq (S^{-1}x, S^{-1}x)(Sy, Sy) = (A^{-1}x,x)(Ay,y)$$

to infer that for any $y \neq 0$ $(A^{-1}x,x) \geqq |(x,y)|^2/(Ay,y)$. However, if we put $y = A^{-1}x$ this inequality becomes an equality. We therefore infer that

$$(A^{-1}x,x) = \sup_{y \neq 0} \frac{|(x,y)|^2}{(Ay,y)}.$$

Clearly, since B is also positive definite, we have

$$(B^{-1}x,x) = \sup_{y \neq 0} \frac{|(x,y)|^2}{(By,y)}.$$

By hypothesis, $(Ay,y) \leqq (By,y)$ for every y, hence $(B^{-1}x,x) \leqq (A^{-1}x,x)$ for all x. Thus $B^{-1} \leqq A^{-1}$. This shows that $-1/\zeta$ belongs to $P_n(0,\infty)$ for all n and a similar argument puts that function in every class $P_n(-\infty,0)$.

An important consequence of the theorem just proved is the following: the class $P_n(a,b)$ does not depend in an essential way on the choice of the interval (a,b). For if (c,d) is another open interval on the real axis, there exists a linear fractional transformation $\varphi(\zeta)$ in the Pick class which maps (c,d) onto (a,b). Hence, if f belongs to $P_n(a,b)$, the composed function $f\circ\varphi$ belongs to $P_n(c,d)$, while if g belongs to $P_n(c,d)$, the function $g\circ\varphi^{-1}$ is in $P_n(a,b)$. In one of our later chapters, it will be convenient to be able to study $P_n(a,b)$ under the hypothesis that $a=-1$ and $b=+\infty$. Our next theorem extends to functions in $P_n(a,b)$ a property that the functions in $P(a,b)$ are known to possess.

Theorem II. *Let f belong to $P_n(a,b)$ and let*

$$a < \xi_1 < \eta_1 < \xi_2 < \eta_2 < \cdots < \xi_n < \eta_n < b$$

be chosen in the interval; then the corresponding Loewner determinant for the function

$$L = \det[\xi_i, \eta_j]$$

is positive.

Proof. As we know, given the system of points ξ_i, η_j in the interval ordered as above, there exists an operator A on the n-dimensional Hilbert space and a suitable one-dimensional projection P on that space so that the operator $B = A + cP$ has exactly the η_j as its spectrum. Here c is an appropriately chosen positive constant. We write all these operators in matrix form, relative to the basis e_k of eigenvectors of A. Thus A appears as a diagonal matrix, while P is the matrix $P_{ij} = \alpha_i \bar{\alpha}_j$, where $\sum |\alpha_i|^2 = 1$. Let $\check{B}$ be the operator having the same eigenvectors as A and defined by the equation $\check{B} e_k = \eta_k e_k$. In matrix form, $\check{B}$ is also a diagonal matrix, and we may write $B = V^{-1} \check{B} V$ where V is a suitable unitary matrix. Now

$$B - A = cP,$$

$$V^{-1} \check{B} V - A = cP$$

and multiplying by V we get

$$\check{B} V - VA = cVP.$$

We now compute the matrix elements for the k-th row and l-th column.

$$(\check{B} V)_{kl} = \sum_j \check{B}_{kj} V_{jl} = \eta_k V_{kl},$$

$$(VA)_{kl} = \sum_j V_{kj} A_{jl} = \xi_l V_{kl},$$

$$(cVP)_{kl} = c \sum_j V_{kj} P_{jl} = c \sum_j V_{kj} \alpha_j \bar{\alpha}_l = c \bar{\alpha}_l F_k$$

where we have put $F_k = \sum_j V_{kj} \alpha_j$.

Evidently

$$(\eta_k - \xi_l) V_{kl} = c \bar{\alpha}_l F_k.$$

We infer that

$$V_{kl} = \frac{c \bar{\alpha}_l F_k}{\eta_k - \xi_l}.$$

Since f belongs to $P_n(a,b)$, $f(B) - f(A) \geqq 0$, and this may be written

$$V^{-1} f(\check{B}) V - f(A) \geqq 0.$$

Hence

$$\det V^{-1} \det[f(\check{B}) V - V f(A)] \geqq 0$$

and this we may write out explicitly in terms of the matrix elements:

$$\det V^{-1} \det \left[\frac{f(\eta_k) - f(\xi_l)}{\eta_k - \xi_l} c \bar{\alpha}_l F_k \right] \geqq 0.$$

Finally

$$\det V^{-1} c^n \prod_{k=1}^{n} \bar{\alpha}_k F_k \det[\eta_i, \xi_l] \geqq 0.$$

Now $\det V^{-1}$ is the reciprocal of $\det V = c^n \prod_{k=1}^{n} \bar{\alpha}_k F_k \det[1/(\eta_i - \xi_l)]$. It follows that the Loewner determinant $\det[\xi_i, \eta_j]$ has the same sign as $\det[1/(\eta_k - \xi_l)]$, and we know the latter determinant to be positive. This completes the proof.

We first apply the previous theorem to the study of a non-constant function $f(x)$ in the class $P_2(a,b)$. In the open interval (a,b) we select

$$\xi_1 < \eta_1 < \xi_2 < \eta_2$$

and find that

$$\det\begin{bmatrix} [\xi_1, \eta_1] & [\xi_1, \eta_2] \\ [\xi_2, \eta_1] & [\xi_2, \eta_2] \end{bmatrix} \geqq 0 .$$

Now hold ξ_1 and η_2 fixed, and permit η_1 and ξ_2 to vary in some closed subinterval $[c,d]$ of (ξ_1, η_2). Let $M = \max[|f(\xi_1)|, |f(\eta_2)|]$; then $[\xi_1, \eta_1]$ is bounded by $2M/(c - \xi_1)$ as η_1 varies over $[c,d]$ and similarly $[\xi_2, \eta_2]$ is bounded by $2M/(\eta_2 - d)$. Accordingly

$$[\xi_1, \eta_2][\xi_2, \eta_1] \leqq \frac{4M^2}{(\eta_2 - d)(c - \xi_1)}$$

as η_1 and ξ_2 vary over $[c,d]$. If $[\xi_1, \eta_2] = 0$, the function is constant over $[c,d]$, otherwise the difference quotients $[\xi_2, \eta_1]$ are uniformly bounded over that interval. Accordingly, $f(x)$ is Lipschitzian over any closed subinterval of the open interval (a,b), and therefore, in particular, is absolutely continuous over any such subinterval.

Now choose two points ξ_1 and ξ_2 in the interval such that the derivative $f'(x)$ exists and is finite at those points. Set $\eta_1 = \xi_1 + \varepsilon$ and $\eta_2 = \xi_2 + \varepsilon$ and let ε tend to 0. We obtain

$$\det\begin{bmatrix} f'(\xi_1) & [\xi_1, \xi_2] \\ [\xi_1, \xi_2] & f'(\xi_2) \end{bmatrix} \geqq 0 .$$

Now if $f'(\xi_1) = 0$, it follows that $[\xi_1, \xi_2] = 0$ for almost every choice of ξ_2 in the interval, and the monotone function $f(x)$ is a constant. We infer that the derivative of a non-constant function in $P_2(a,b)$ never vanishes.

Next, suppose that $f(x)$ in $P_2(a,b)$ is non-constant and smooth, say in the class C^3. We write the determinant of the extended Loewner matrix

$$\det\begin{bmatrix} [\xi_1, \eta_1] & [\xi_1, \eta_1, \eta_2] \\ [\xi_1, \xi_2, \eta_1] & [\xi_1, \xi_2, \eta_1, \eta_2] \end{bmatrix}$$

which we know to be non-negative. Let the points ξ_i, η_j all converge to some fixed point x in the interval; we have in the limit

$$\det M_2(x,f) = \det \begin{bmatrix} f'(x) & \dfrac{f''(x)}{2!} \\ \dfrac{f''(x)}{2!} & \dfrac{f^{(3)}(x)}{3!} \end{bmatrix} \geqq 0 .$$

Since f is non-constant, $f'(x)$ is strictly positive, hence $f^{(3)}(x)$ can never be negative. It follows that $f'(x)$ is a positive, convex function.

We can now establish the following theorem, describing the local properties of functions in $P_2(a,b)$.

Theorem III. *Let $f(x)$ belong to $P_2(a,b)$; then*
(i) *$f(x)$ is continuously differentiable,*
(ii) *the derivative is convex and non-negative, and*
(iii) *the matrix function*

$$M_2(x,f) = \begin{bmatrix} f'(x) & \dfrac{f''(x)}{2!} \\ \dfrac{f''(x)}{2!} & \dfrac{f^{(3)}(x)}{3!} \end{bmatrix}$$

(which makes sense almost everywhere) is a positive matrix almost everywhere.

Proof. If $f(x)$ is a constant, the assertions of the theorem are trivially true; we may therefore suppose $f(x)$ not a constant, and therefore, that $f'(x)$ is strictly positive where it exists. Since the function is monotone, that derivative exists almost everywhere.

Let $f_\varepsilon(x)$ be the regularization of f of order ε; this is a smooth function in $P_2(a+\varepsilon, b-\varepsilon)$ and our theorem has already been proved for smooth functions. Since $f(x)$ is absolutely continuous, we know that the derivatives $(d/dx) f_\varepsilon(x)$ are the regularizations of $f'(x)$, and these derivatives converge with decreasing ε to $f'(x)$ almost everywhere. However, those derivatives form a family of convex functions, and as we shall see, they are equicontinuous on any closed subinterval of (a,b). Let $[c,d]$ be such a closed subinterval and x and y two points of the subinterval; now, from the Mean Value Theorem

$$\left| \frac{f'_\varepsilon(x) - f'_\varepsilon(y)}{x-y} \right| = |f''_\varepsilon(\xi)| \leqq \operatorname{Max}(|f''_\varepsilon(c)|, |f''_\varepsilon(d)|)$$

$$\leqq \operatorname{Max}\left(\frac{|f'_\varepsilon(c) - f'_\varepsilon(c-h)|}{h}, \frac{|f'_\varepsilon(d+h) - f'_\varepsilon(d)|}{h} \right).$$

If we select c, d and h so that the system $f'_\varepsilon(x)$ converges at the four points $c-h$, c, d and $d+h$, then the family is certainly equicontinuous on $[c,d]$. Because that convergence takes place almost everywhere, the family is uniformly Lipschitzian on any closed subinterval of (a,b). This circumstance makes it clear that the convex and positive functions $f'_\varepsilon(x)$ converge uniformly on closed subintervals of (a,b) to a convex and positive limit. Thus, $f(x)$ is the indefinite integral of a convex function. This establishes (i) and (ii). Since $f'(x)$ is convex, it is surely absolutely continuous on closed subintervals and therefore the second derivatives $d^2/dx^2 f_\varepsilon(x)$ are the regularizations of the monotone function $f''(x)$. Those regularizations converge to $f''(x)$ with decreasing ε at all points of continuity of $f''(x)$. A theorem proved in Chapter I now guarantees that the derivatives $f_\varepsilon^{(3)}(x)$ converge to $f^{(3)}(x)$ at all x where that derivative exists and is finite. Hence, almost everywhere, $M_2(x,f_\varepsilon)$ converges to $M_2(x,f)$, which is therefore a positive matrix too. This completes the proof of the theorem.

Suppose, again, that $f(x)$ is a smooth function in $P_2(a,b)$. The convexity of $f'(x)$ is expressed by the positivity of $f^{(3)}(x)$. The logarithmic convexity of the function is expressed by the inequality $f''(x)^2 \leqq f'(x)f^{(3)}(x)$. However, the inequality which we have found is even stronger: $f''(x)^2 \leqq (2/3)f'(x)f^{(3)}(x)$. It is worthwhile to establish the geometric meaning of this inequality. We shall suppose $f(x)$ non-constant, so that $f'(x)$ is convex and strictly positive. We may therefore take the positive square root of its reciprocal. We put

$$f'(x) = \frac{1}{c(x)^2}$$

where $c(x)$ is positive. The inequality which we have established then shows that $c(x)$ must be concave. Indeed, putting

$$c(x) = (f'(x))^{-\frac{1}{2}}$$

we have

$$c''(x) = (f'(x))^{-\frac{5}{2}} \tfrac{3}{4}\left[f''(x)^2 - \tfrac{2}{3} f'(x) f^{(3)}(x)\right].$$

Since the convex $f'(x)$ is the limit of its regularizations everywhere, we obtain the following theorem, which does not require that the function be smooth.

Theorem IV. *If $f(x)$ is not constant, and belongs to $P_2(a,b)$, it is then the indefinite integral of a function of the form $1/c(x)^2$, where $c(x)$ is positive and concave on the interval.*

The following theorem is one which we already know for functions in the class $P(a,b)$; we are now able to extend it to $P_n(a,b)$.

Theorem V. *Let $f(x)$ belong to $P_n(a,b)$ for $n \geqq 2$ and $\lambda_1, \lambda_2, \ldots, \lambda_n$ be n distinct points of the open interval (a,b); then the Pick matrix formed with the divided differences of f*

$$K_{ij} = [\lambda_i, \lambda_j]$$

is a positive matrix.

Proof. A permutation of the λ's merely corresponds to a permutation of the basis elements relative to which the matrix K represents a transformation; there is therefore no loss of generality in assuming $\lambda_1 < \lambda_2 < \cdots < \lambda_n$. We may also suppose that f is not a constant, since in that case K is the 0 matrix. It follows that the diagonal elements are strictly positive, since f is surely in $P_2(a,b)$. The theorem has already been proved for $n=2$, and we prove it for higher values of n by induction. The passage from n to $n+1$ is carried out in the following way. Let $\varphi(\zeta)$ be a Pick function in the class $P(a,b)$ which is not a rational function, and let K'_{ij} be the corresponding Pick matrix determined by the λ's. From a theorem in Chapter III we know K' to be a positive definite matrix. For small positive values of ε the function $f(x)+\varepsilon\varphi(x)$ is in $P_n(a,b)$ and is associated with the Pick matrix

$$K''_{ij} = K_{ij} + \varepsilon K'_{ij}.$$

We use the criterion of Chapter I to show that this is a positive matrix. The square matrix of order n obtained by deleting the final row and column of K_{ij} is a positive matrix by the inductive hypothesis. The corresponding matrix obtained from K'_{ij} is even positive definite, since $\varphi(\zeta)$ was not a rational function. Thus, the matrix obtained from K'' by deleting the final row and column is a positive definite matrix. Hence the first n determinants in the sequence of subdeterminants of K'' are in fact strictly positive. However, the determinant of K'' is also positive, since that determinant must be a polynomial of degree $n+1$ in ε; the polynomial is not a constant, since the coefficient of the leading term is $\det K'$, and this is strictly positive. The polynomial itself is positive for $\varepsilon > 0$ since $\det K''_{ij}$ is the limit as h tends to 0 of determinants of the form

$$\det \begin{bmatrix} [\lambda_1, \lambda_1+h] & [\lambda_1, \lambda_2+h] & \ldots & [\lambda_1, \lambda_n+h] \\ [\lambda_2, \lambda_1+h] & [\lambda_2, \lambda_2+h] & \ldots & [\lambda_2, \lambda_n+h] \\ \vdots & \vdots & & \vdots \\ [\lambda_n, \lambda_1+h] & [\lambda_n, \lambda_2+h] & \ldots & [\lambda_n, \lambda_n+h] \end{bmatrix}$$

and these determinants are non-negative in view of the fact that $f(x)+\varepsilon\varphi(x)$ is in $P_n(a,b)$. Thus K'' is a positive matrix and as ε approaches 0, K is the limit of a sequence of positive matrices. This completes the proof.

We can also describe the local properties of functions in the class $P_n(a,b)$ in a now familiar way, if we make use of the matrix function $M_n(x,f)$ defined as follows:

$$M_n(x,f) = \begin{bmatrix} f'(x) & \frac{f''(x)}{2!} & \frac{f^{(3)}(x)}{3!} & \cdots & \frac{f^{(n)}(x)}{n!} \\ \frac{f''(x)}{2!} & \frac{f^{(3)}(x)}{3!} & & \cdots & \frac{f^{(n+1)}(x)}{(n+1)!} \\ \frac{f^{(3)}(x)}{3!} & & & & \\ \vdots & \vdots & \vdots & & \vdots \\ \frac{f^{(n)}(x)}{n!} & & & \cdots & \frac{f^{(2n-1)}(x)}{(2n-1)!} \end{bmatrix}$$

Theorem VI. *Let f belong to the class $P_n(a,b)$ for $n \geqq 2$; then*

(i) *f belongs to C^{2n-3},*

(ii) *the derivative $f^{(2n-3)}(x)$ is convex, and,*

(iii) *the matrix function $M_n(x,f)$, which makes sense almost everywhere, is a positive matrix almost everywhere.*

Proof. We first remark that if $f^{(2n-3)}(x)$ is convex, its derivative exists and is finite, except for at most a countable number of points. Moreover, that derivative is monotone increasing, and hence differentiable almost everywhere. Thus $M_n(x,f)$ exists almost everywhere.

The theorem is already known for $n=2$, and our proof is by induction. We suppose therefore that the theorem is true for n, and pass now to $n+1$. The theorem is first proved for smooth functions. Let $\varphi(\zeta)$ be a function in $P(a,b)$ which is not rational: the matrix $M_n(x,\varphi)$ is then positive definite, in view of a theorem in Chapter III. Accordingly, by virtue of the inductive hypothesis, $M_k(x,f+\varepsilon\varphi)$ is a positive definite matrix when $\varepsilon>0$ for all $k \leqq n$, and therefore $\det M_k(x,f+\varepsilon\varphi)$ is strictly positive. If we next consider $\det M_{n+1}(x,f+\varepsilon\varphi)$ we can show that this determinant is also positive. The determinant is the limit of the extended Loewner determinant computed with points

$$a<\xi_1<\eta_1<\xi_2<\eta_2<\cdots<\xi_{n+1}<\eta_{n+1}<b$$

when these points converge to the fixed value x. Thus the determinant is non-negative. We know that the determinant is strictly positive, since it is a polynomial in ε with leading coefficient $M_{n+1}(x,\varphi)$. We infer that for small positive ε the determinant of $M_{n+1}(x,f+\varepsilon\varphi)$ is strictly positive. This, in turn, makes $M_{n+1}(x,f+\varepsilon\varphi)$ a positive matrix, which may be written $M_{n+1}(x,f)+\varepsilon M_{n+1}(x,\varphi)$. As ε tends to 0, this approaches the limit $M_{n+1}(x,f)$, which is therefore a positive matrix, as desired. We deduce that $f^{(2n-1)}(x)$ is non-negative and convex, since its second derivative is non-negative.

Now let $f(x)$ be an arbitrary element of the class $P_{n+1}(a,b)$; we pass to its regularizations f_ε. Since these are smooth functions, we know that $f_\varepsilon^{(2n-1)}(x)$ is positive and convex, as well as that $M_{n+1}(x,f_\varepsilon)$ is a positive matrix for all x in the interval $(a+\varepsilon, b-\varepsilon)$.

Consider now the monotone increasing function $f^{(2n-2)}(x)$. As yet, we do not know that this function is absolutely continuous, or even continuous. However, its regularizations converge to $f^{(2n-2)}(x)$ at every point of continuity, which means outside of a set which is at most countable. By a Theorem in Chapter I, the derivatives $f_\varepsilon^{(2n-1)}$ converge to the derivative $f^{(2n-1)}(x)$ whenever that derivative exists and is finite, hence almost everywhere. Since the $f_\varepsilon^{(2n-1)}(x)$ form a family of convex functions, converging almost everywhere on the interval, they are in fact equicontinuous on closed subintervals, and so the convergence is uniform on closed subintervals. This in turn implies that $f^{(2n-1)}(x)$ is also convex, whence $f^{(2n-2)}(x)$ appears as the indefinite integral of a convex function, plus, possibly, a singular function, i.e., one with derivative 0 almost everywhere. We must show that no such singular function exists, thereby putting f in the class C^{2n-1}. For this purpose we simply note that if a singular term existed in the canonical expression for $f^{(2n-2)}(x)$ as a monotone function, there would exist values $h>0$ and points c, $c+h$ in the interval for which the difference quotients $(f^{(2n-2)}(c+h)-f^{(2n-2)}(c))/h$ would be arbitrarily large, the points c, $c+h$ also being points of continuity of the function. Since this difference quotient can be approximated by difference quotients of the regularizations, we find that difference quotients of the regularizations are arbitrarily large over the interval. However, an estimate which we have already given when we proved this theorem in the special case $n=2$ shows that such difference quotients are bounded over closed subintervals. Thus (i) and (ii) are established. To prove (iii) we have only to remark as before that the first and second derivatives of the regularizations of the convex $f^{(2n-1)}$ converge pointwise almost everywhere to the corresponding derivatives of $f^{(2n-1)}$, whence $M_{n+1}(x,f)$ is almost everywhere the pointwise limit of a sequence of positive matrices. This completes the proof.

Chapter VIII. Sufficient Conditions

In the previous chapter we introduced the classes $P_n(a,b)$ and derived several conditions which functions in $P_n(a,b)$ must necessarily satisfy. In this chapter we will show that many of these conditions are also sufficient. Of course, we are always tacitly supposing that $n \geqq 2$, since the case $n=1$ is a trivial one.

We should first fix our attention on the functions to be studied as operator functions. For this purpose, we consider more carefully the space of all symmetric operators on an n-dimensional Hilbert space. This is a vector space over the reals with dimension n^2. Since the operator space is finite dimensional, there is only one natural topology, although there exist a variety of norms, all, of course, equivalent. We shall find it convenient to take the Hilbert-Schmidt norm on the operator space, defined as follows:

$$\|H\|^2 = \mathrm{Trace}(HH^*)\,.$$

Of course, since the operators we consider are all symmetric, this could be written

$$\|H\|^2 = \mathrm{Trace}(H^2) = \sum_{i=1}^{n} \lambda_i^2$$

where the λ_i are the eigenvalues of H. However, it is not necessary to have H in diagonal form to be able to compute the norm: if H is written as a matrix H_{ij} relative to any fixed orthonormal basis, then $\|H\|^2 = \sum_{i,j} H_{ij}^2$. Under this quadratic norm the operator space itself becomes a Hilbert space, however, we will not have occasion to use this fact in the sequel.

Now we consider a real function $f(x)$ defined on the open interval (a,b). Corresponding to that interval is an open set in the operator space, consisting of all symmetric operators with spectra in (a,b). Our function $f(x)$ now gives rise to an operator function defined on that open set. That function, as a mapping from an open subset of the n^2-dimensional operator space into itself often has a differential, which we now seek to identify.

Let A be an operator in the domain of the operator function f. That function has a differential at A if there exists a linear transformation $L_{f,A}$ of the operator space into itself such that for all H

$$\|f(A+H)-f(A)-L_{f,A}(H)\| = o(\|H\|).$$

The important result is the following theorem.

Theorem I. *Let $f(x)$ be a continuously differentiable real function on the interval (a,b); then the corresponding operator function has a differential at every operator A in its domain. To determine $L_{f,A}(H)$, write every operator as a matrix in terms of the orthonormal basis of eigenvectors of A; then $L_{f,A}(H)$ is the Schur product $K \circ H$ where K is the Pick matrix associated with the function $f(x)$ and the eigenvalues λ_i of A.*

Proof. Consider, first, the special case of a function $f(x)$ such that $f(\lambda_i)=f'(\lambda_i)=0$ for all eigenvalues λ_i of the operator A. Evidently, we want to show that this operator function has the differential 0 at A. Now

$$f(A+H)-f(A)=f(A+H)$$

so it is necessary to show that

$$\|f(A+H)\| \quad \text{is} \quad o(\|H\|).$$

When the norm of H is small, the eigenvalues of $A+H$ are very close to those of A, and so we may write them $\lambda_i+\varepsilon_i$ where we do not suppose the ε_i necessarily positive. If $\|H\|=\varepsilon$ then $|\varepsilon_i| \leqq \sqrt{n}\,\varepsilon$ and we find that

$$\|f(A+H)\|^2 = \sum |f(\lambda_i+\varepsilon_i)|^2 = \sum |f(\lambda_i+\varepsilon_i)-f(\lambda_i)|^2$$

and by virtue of the Mean Value theorem this becomes

$$\sum |f'(\lambda_i+\varepsilon_i')|^2\, \varepsilon_i^2 \leqq n\,\varepsilon^2 \sum |f'(\lambda_i+\varepsilon_i')|^2 .$$

The coefficient of ε^2 diminishes to 0 as ε does, and so $\|f(A+H)\|$ is $o(\varepsilon)$, as desired.

More generally, then, given a continuously differentiable $f(x)$ we construct a polynomial $p(x)$ which coincides with the function at the eigenvalues of A and such that the derivative $p'(x)$ coincides with $f'(x)$ at those same points. The difference, $f(x)-p(x)$, then vanishes, together with its derivative, at those points. Accordingly, the difference, considered as an operator function, has the differential 0 at A. It follows that f has a differential at A if and only if the polynomial p does, and the differential is the same. We should emphasize here that the Pick matrix associated with f at the eigenvalues of A coincides with the Pick matrix corresponding to the polynomial $p(x)$. It therefore follows that we need only prove the theorem for polynomials. This we do by induction on the degree of the polynomial.

Suppose $f(x)$ is the linear function $f(x)=\alpha x+\beta$. For the operator function we may write

$$f(A+H)-f(A)=\alpha H$$

to see that

$$\|f(A+H)-f(A)-\alpha H\|=0.$$

Thus the differential $L_{f,A}$ is simply the linear transformation of multiplication by the constant α. However, since the Pick matrix associated with $f(x)$ is simply αE, the matrix having the entry α in every place, it is clear that our theorem holds for polynomials of degree 0 or 1.

In carrying out the inductive argument, we are at liberty to suppose that the polynomials in question vanish at the origin. Let $p(x)$ be of degree $n+1$ and suppose the theorem proved for n and that $p(x)=x\,q(x)$. Now

$$\begin{aligned} p(A+H)-p(A) &= (A+H)q(A+H)-A\,q(A) \\ &= (A+H)(q(A+H)-q(A))+H\,q(A) \\ &= (A+H)\,L_{q,A}(H)+H\,q(A)+Z \end{aligned}$$

where $\|Z\|=o(\|H\|)$. Since $H\,L_{q,A}(H)$ obviously satisfies a similar inequality, the differential $L_{p,A}(H)$ is given by $A\,L_{q,A}(H)+H\,q(A)$. Now we suppose that all these operators are written as matrices relative to the orthonormal basis of eigenvectors of A. We compute the matrix elements for the k-th row and l-th column as follows.

$$\begin{aligned} (L_{q,A}(H))_{kl} &= [\lambda_k,\lambda_l]_q\,H_{kl}, \\ (A\,L_{q,A}(H))_{kl} &= \lambda_k[\lambda_k,\lambda_l]_q\,H_{kl}, \\ (H\,q(A))_{kl} &= q(\lambda_l)\,H_{kl}. \end{aligned}$$

For the differential $L_{p,A}(H)$ in the k-th row and l-th column, then we get

$$\begin{aligned} (L_{p,A}(H))_{kl} &= \{\lambda_k[\lambda_k,\lambda_l]_q+q(\lambda_l)\}\,H_{kl} \\ &= [\lambda_k,\lambda_l]_p\,H_{kl} \end{aligned}$$

and this evidently the Schur product of the Pick matrix associated with $p(x)$ and the matrix representing H relative to the eigenvectors of A. Thus the proof is complete.

The previous theorem enables us to establish a sufficient condition for a function to belong to $P_n(a,b)$.

Theorem II. *Let $f(x)$ be a real, C^1 function on the interval (a,b) such that for any n points $\lambda_1,\lambda_2,\ldots,\lambda_n$ of that interval, the corresponding Pick matrix*

$$K_{ij}=[\lambda_i,\lambda_j]_f$$

is a positive matrix; then f belongs to $P_n(a,b)$.

Proof. Let A and B be arbitrary symmetric operators on an n-dimensional Hilbert space having their spectra in the interval (a,b); we also suppose that $A \leqq B$, and have to show that $f(A) \leqq f(B)$. Put $H = B - A$ and form

$$A_t = A + tH \quad \text{for } 0 \leqq t \leqq 1 .$$

Let u be an arbitrary vector of the Hilbert space and let

$$F(t) = (f(A_t)u, u) .$$

Evidently $F(0) = (f(A)u,u)$ and $F(1) = (f(B)u,u)$, and so it will be enough to show that $F(t)$ is always a monotone non-decreasing function. In view of the previous theorem, we know that $F(t)$ is differentiable and we can even compute the derivative:

$$F'(t) = (K \circ H u, u) .$$

Because the Schur product of positive matrices is again a positive matrix we have $F'(t) \geqq 0$ and therefore $F(t)$ is monotone. Hence $f(x)$ is in $P_n(a,b)$.

It is advisable to recall at this point that in our study of necessary conditions for a function to belong to $P_n(a,b)$, the condition that the associated Pick matrix be positive was derived from the positivity of the Loewner determinant computed with an arbitrary system of the form

$$a < \xi_1 < \eta_1 < \cdots < \xi_n < \eta_n < b ;$$

Accordingly, although it is unnecessary to state it as a formal theorem, the knowledge, for a function on (a,b) that all such Loewner determinants are positive, is sufficient to ensure that the function is in $P_n(a,b)$.

We now find that the description which has been given of functions in the class $P_2(a,b)$ is complete.

Theorem III. *Let $f(x)$ be the indefinite integral of a function of the form $1/c(x)^2$, where $c(x)$ is positive and concave on the interval (a,b); then $f(x)$ belongs to $P_2(a,b)$.*

Proof. In view of the previous theorem, since $f(x)$ is real and C^1 all that is required is to show that the Pick matrix associated with any two points λ_1, λ_2 of (a,b)

$$K = \begin{bmatrix} [\lambda_1, \lambda_1] & [\lambda_1, \lambda_2] \\ [\lambda_2, \lambda_1] & [\lambda_2, \lambda_2] \end{bmatrix}$$

is a positive matrix. Now we know that the diagonal elements of this symmetric matrix are strictly positive, so it is only necessary to show that the determinant is positive. This reduces to the inequality

$$\left|\frac{1}{\lambda_2-\lambda_1}\int_{\lambda_1}^{\lambda_2}\frac{dt}{c(t)^2}\right|^2 \leqq \left|\frac{1}{c(\lambda_1)c(\lambda_2)}\right|^2$$

or, supposing, as we may, that $\lambda_1 < \lambda_2$,

$$\frac{1}{\lambda_2-\lambda_1}\int_{\lambda_1}^{\lambda_2}\frac{c(\lambda_1)c(\lambda_2)}{c(t)^2}\,dt \leqq 1\,.$$

We make the change of variables: $s=(t-\lambda_1)/(\lambda_2-\lambda_1)$ and put $k(s)=c(s\lambda_2+(1-s)\lambda_1)$ to obtain

$$\int_0^1 \frac{k(0)k(1)}{k(s)^2}\,ds \leqq 1\,.$$

Since $k(s)$ is concave, $k(s) \geqq s\,k(1)+(1-s)\,k(0)$, accordingly

$$\int_0^1 \frac{k(0)k(1)}{k(s)^2}\,ds \leqq \int_0^1 \frac{k(0)k(1)}{(s\,k(1)+(1-s)\,k(0))^2}\,ds = 1\,.$$

Thus the proof is complete.

The theorem just proved could be put in an equivalent form as follows.

Theorem IV. *Let $f(x)$ be a C^1 function on the interval (a,b) for which $f'(x)$ is positive and convex; if in addition the matrix function*

$$M_2(x,f) = \begin{bmatrix} f'(x) & \dfrac{f''(x)}{2!} \\[2ex] \dfrac{f''(x)}{2!} & \dfrac{f^{(3)}(x)}{3!} \end{bmatrix}$$

is a positive matrix almost everywhere, then $f(x)$ belongs to $P_2(a,b)$.

It now becomes clear that the class P_2 is a local property. This means that if two intervals (a,b) and (c,d) overlap, say

$$a<c<b<d$$

and a function $f(x)$, defined on (a,d) is in $P_2(a,b)$ as well as $P_2(c,d)$, then $f(x)$ belongs to $P_2(a,d)$. The easy proof goes as follows: if $f(x)$ is a constant, there is nothing to be shown, while if the function is not a constant, then its derivative is the reciprocal of a positive function, concave on (a,b) as well as concave on (c,d). Such a function is evidently concave on the union of those two intervals, and hence $f(x)$ is in $P_2(a,d)$. This result holds for all values of n, and not simply $n=1$ and $n=2$, but the proof for higher values is difficult and will depend on the theory of the Cauchy Interpolation Problem, to be considered in a later chapter. Hence, although we postpone the proof, we state the theorem.

Theorem. *For all values of n, P_n is a local property.*

We make use of this theorem to complete our description of the classes $P_n(a,b)$.

Theorem V. *Let $f(x)$ be a real function on (a,b) which belongs to the class C^{2n-3} and such that the $2n-3rd$ derivative is positive and convex; then $f(x)$ belongs to $P_n(a,b)$ if the corresponding matrix function $M_n(x,f)$ is almost everywhere a positive matrix.*

Proof. We first remark that we can suppose that the function is a smooth function, since the regularizations of $f(x)$ will satisfy the hypotheses of the theorem, and these regularizations converge pointwise to $f(x)$. Since $P_n(a,b)$ is closed under pointwise convergence, it follows that $f(x)$ itself is in that class.

One point in the foregoing argument must be considered more carefully. If the matrix function $M_n(x,f)$ is regularized, the resulting smooth matrix function is not exactly $M_n(x,f_\varepsilon)$. The two matrices may differ in the entry in the lower right hand corner. This will happen when $f^{(2n-2)}(x) = m(x)+S(x)$ where $m(x)$ is absolutely continuous, and $S(x)$ a singular, monotone function. Then $f^{(2n-1)}(x)=m'(x)$. Thus the regularization of the matrix function will have $m_\varepsilon(x)+S_\varepsilon(x)$ in the position corresponding to $f^{(2n-2)}$, as well as $m'_\varepsilon(x)$ in the corner. On the other hand, the $2n-2nd$ derivative of the regularized function will be $m_\varepsilon(x)+S_\varepsilon(x)$, as before, but in the corner will occur the $m'_\varepsilon(x)+S'_\varepsilon(x)$. Thus the corner term of $M_n(x,f_\varepsilon)$ will be at least as large, and often larger, than that of the regularizations, or averages, of $M_n(x,f)$. Thus $M_n(x,f_\varepsilon)$ is surely a positive matrix for all x if $M_n(x,f)$ is almost everywhere a positive matrix.

We may also suppose that the smooth function $f(x)$ gives rise to a positive definite $M_n(x,f)$ over the interval (a,b). For this purpose we have only to add to $f(x)$ a suitable function of the form $\varepsilon\varphi(x)$ where ε is small and positive and $\varphi(x)$ is a function in the class $P(a,b)$ which is not rational. Now, again, $f(x)$ is the pointwise limit of $f(x)+\varepsilon\varphi(x)$, and if the latter functions are in $P_n(a,b)$, so is $f(x)$. But it is clear that $M_n(x,f+\varepsilon\varphi)=M_n(x,f)+\varepsilon M_n(x,\varphi)$ makes $M_n(x,f+\varepsilon\varphi)$ positive definite.

In particular, then, the smallest eigenvalue of $M_n(x,f)$ is a function $\lambda(x)$ which is continuous on the interval (a,b) and always positive there.

Choose a point x_0 in the interval. We assert that there exists a positive h so that $f(x)$ is in the class $P_n(x_0-h, x_0+h)$. For if this were not the case, then for every small h there would exist a system of $2n$ numbers so ordered that

$$x_0-h<\xi_1<\eta_1<\xi_2<\eta_2<\cdots<\xi_n<\eta_n<x_0+h$$

and such that the corresponding Loewner determinant would be negative. It follows that the corresponding extended Loewner determinant would also be negative. However, as h tends to 0, these extended Loewner matrices would converge pointwise to $M_n(x_0, f)$, and the determinant of this matrix is strictly positive, indeed $\geqq \lambda(x_0)^n$.

Let $[c,d]$ be any closed subinterval of (a,b). This interval is covered by a family of open intervals, on each of which the function belongs to the class P_n; by virtue of the compactness of $[c,d]$ finitely many such subintervals cover $[c,d]$, and in view of the fact that P_n is a local property, it follows that f belongs to $P_n(c,d)$. Because c and d were arbitrary in (a,b), f is in $P_n(a,b)$ as desired.

We see that the necessary conditions derived in the previous chapter for a function to be in $P_n(a,b)$ are also sufficient, whether the conditions refer to the Loewner determinants, the Pick matrices, or the matrix functions $M_n(x,f)$. Unfortunately, we have had to invoke the theorem that P_n is a local property without having given its proof. In the absence of that theorem, we would not yet be sure that the classes $P_n(a,b)$ were all distinct as n increases. However, in the next chapter, we shall investigate the intersection of the classes $P_n(a,b)$ and shall find that this intersection consists exactly of the functions in $P(a,b)$.

Chapter IX. Loewner's Theorem

This chapter is devoted to a proof of the following theorem due to Loewner.

Theorem. *Let $f(x)$ belong to $P_n(a,b)$ for every value of n; then f belongs to the class $P(a,b)$.*

This remarkable theorem should perhaps be regarded as a theorem about analytic continuation: a function $f(x)$ on an interval (a,b) satisfies a certain infinite family of inequalities; we infer that it can be continued analytically over the whole upper and lower half-planes. No such theorem can be easy to prove, and Loewner's theorem is not an elementary one. Three proofs have been given. Loewner's original proof will appear later in this book, and makes use of the study of the Cauchy Interpolation Problem. An interesting proof has been given by Koranyi and Nagy, whose method depends on some facts concerning the theory of self-adjoint operators in a Hilbert space, as well as the theory of reproducing kernels. That proof also will appear in a subsequent chapter. Doubtless the simplest proof is essentially due to Bendat and Sherman, and their argument is followed in this chapter. The only feature of this argument which is not elementary is the invocation of the theory of the Hamburger Moment Problem. Thus, in the sequel, we shall need the following result.

Theorem. *Let c_k, $k=0,1,\ldots$ be a sequence of real numbers such that for every n the matrices*

$$C_n = \begin{bmatrix} c_0 & c_1 & c_2 & \ldots & c_n \\ c_1 & c_2 & & \ldots & c_{n+1} \\ c_2 & & & & \\ \vdots & \vdots & & & \vdots \\ c_n & c_{n+1} & & \ldots & c_{2n} \end{bmatrix}$$

are positive matrices; then there exists a positive Radon measure ν on the real axis such that for every k

$$c_k = \int \lambda^k \, d\nu(\lambda).$$

We proceed with the proof of Loewner's theorem. As we have earlier remarked, it is legitimate for us to assume that the interval (a,b) in question is the infinite interval $(-1,+\infty)$. We are also at liberty to assume that the function $f(x)$ is bounded, since otherwise we can compose f with a suitable linear fractional transformation g in the Pick class to obtain a bounded $g\circ f$ which is also in every $P_n(-1,+\infty)$. If the composed function is a Pick function, so is the original $f(x)$. Because $f(x)$ belongs to every $P_n(-1,\infty)$ the function is C^∞, and because the function is bounded and increasing, its derivative $f'(x)$ converges to 0 with increasing x.

We first show that $f'(x)$ is completely monotone on the interval $(-1,+\infty)$. We know that all of the odd derivatives of $f(x)$ are convex and positive, and we must show that the even derivatives of the function are negative. This is the same thing as showing that all of the odd derivatives, convex and positive, are decreasing functions of x. This, in turn, is a consequence of the following lemma.

Lemma. *Let $g(x)$ be a smooth function on $(-1,+\infty)$ which is positive, convex and decreasing, and with a convex second derivative: then that second derivative is also decreasing.*

Proof. If the second derivative were ever increasing, then $g'(x)$, for large x, would be positive, contradicting the fact that $g(x)$ is decreasing. Accordingly, from the convexity of $f'(x)$ and $f^{(3)}(x)$, as well as the fact that $f'(x)$ diminishes to 0, we infer that $f^{(3)}(x)$ also does. We argue inductively through the sequence of odd derivatives to infer that all of the odd derivatives decrease with increasing x, hence that all of the even derivatives are negative. So, putting $g(x)=f'(x)$ we have

$$(-1)^n g^{(n)}(x)\geqq 0$$

for all n and all $x>-1$. Thus $f'(x)$ is completely monotone, and admits an analytic continuation to the half-plane $\mathrm{Re}[\zeta]=\xi>-1$ in view of the Little Bernstein Theorem. For any point in that half-plane, then

$$f(z)=f(0)+\int_0^z g(\zeta)\,d\zeta$$

where the integration is taken over a line segment from the origin to z. Now f occurs as the restriction to the interval of a function analytic in a half-plane.

The function $f(z)$ may now be expanded in a McLaurin series:

$$f(z)=a_0+a_1 z+a_2 z^2+\cdots$$

convergent in the unit disk $|z|<1$. Because $f(x)$ is in every P_n class, the matrices

$$M_n(0,f)=\begin{bmatrix} a_1 & a_2 & a_3 \dots a_n \\ a_2 & a_3 & \\ \vdots & & \vdots \\ a_n & & \dots a_{2n-1} \end{bmatrix}$$

are all positive matrices. Hence, from the theory of the Hamburger moment problem, there exists a positive Radon measure ν on the real axis such that

$$a_k=\int \lambda^{k-1}\, d\nu(\lambda)$$

for all k. We may therefore write the series for $f(z)$ as follows:

$$f(z)=a_0+\sum_{k=1}^{\infty}\int \lambda^{k-1}\, d\nu(\lambda)\, z^k .$$

Since the series is convergent for $|z|<1$ we must have

$$\limsup_{k\to\infty} |a_k|^{1/k}\leqq 1 .$$

From this it follows that the measure ν is supported by the closed interval $[-1,1]$ for otherwise, if ν had a mass $m>0$ outside an interval of the form $(-1-\varepsilon, 1+\varepsilon)$ where $\varepsilon>0$ then

$$\int \lambda^{2k}\, d\nu(\lambda)\geqq m(1+\varepsilon)^{2k}$$

and therefore $\limsup_{k\to\infty}|a_k|^{1/k}\geqq 1+\varepsilon$, a contradiction. It follows that the summation and integration may be interchanged, provided $|z|<1$ to obtain

$$f(z)=a_0+\int \frac{z}{1-z\lambda}\, d\nu(\lambda) .$$

It is easy to check that the integral makes sense and is analytic whenever z is not real, and therefore provides an analytic continuation of $f(z)$ outside the disk $|z|<1$ throughout the open upper and lower half-planes. It is almost obvious that the function so obtained is a Pick function. Of course the measure ν occuring in the formula above is not the canonical measure associated with that Pick function. However, the proof is now complete.

Chapter X. Reproducing Kernels

A function $K(x,y)$ defined on a product space $S \times S$ is called a *positive matrix* if, for any finite set

$$x_1, x_2, \ldots, x_n$$

of points of S and equally many complex numbers

$$z_1, z_2, \ldots, z_n$$

the inequality

$$\sum \sum K(x_i, x_j) z_i \bar{z}_j \geqq 0$$

holds. It is clear that the definition could equally well require that the matrix

$$K_{ij} = K(x_i, x_j)$$

associated with any finite subset of S be a positive matrix in the sense of Chapter I.

The set S being given, a positive matrix may be generated on $S \times S$ in the following way. We consider a mapping

$$x \to K_x$$

of S into some Hilbert space $\mathscr{H}$ and form the function

$$K(x,y) = (K_y, K_x).$$

It is clear that the restriction of this function to $F \times F$ where F is a finite subset of S is a Gram's matrix, hence a positive matrix. It is important for us to note that every positive matrix can be obtained in this way, and that there exists a canonical mapping of S into a canonically associated Hilbert space called the *reproducing kernel* space. We proceed now to the construction of that space and that mapping.

Let $\mathscr{H}_0$ be the linear space consisting of all functions on S which are finite sums of the form

$$u(x) = \sum a_k K(x, y_k).$$

If $v(x)=\sum b_j K(x,z_j)$ is another such function, the inner product of u and v is defined by the formula

$$(u,v)=\sum\sum a_k \bar{b}_j K(z_j,y_k).$$

This number may be written in the equivalent forms

$$\sum a_k \overline{v(y_k)} \quad \text{and} \quad \sum u(z_j)\bar{b}_j$$

and this circumstance shows that the inner product so defined depends only on the values of the functions u and v on the space S and not on the representation of such functions as finite sums of sections of the function $K(x,y)$. Let $K_y(x)$ be the function in $\mathscr{H}_0$ given by $K_y(x)=K(x,y)$. It is easy to see that then, for all u in the space

$$u(x)=(u,K_x).$$

The inner product gives rise to a quadratic form $\|u\|$ defined by

$$\|u\|^2=(u,u)=\sum\sum a_k \bar{a}_j K(y_j,y_k)$$

which is $\geqq 0$ by hypothesis. If u is a function for which $\|u\|=0$ then for every point y of S $|u(y)|=|(u,K_y)|\leqq\|u\|\,\|K_y\|=0$ and the function vanishes identically. Conversely, if the function $u(x)=\sum a_k K(x,y_k)$ vanishes identically, then $(u,u)=\sum u(y_k)\bar{a}_k=0$ and $\|u\|=0$. It follows that $\|u\|$ is in fact a norm on the space $\mathscr{H}_0$ which now appears as a pre-Hilbert space, since there is no a priori reason for supposing that it is complete. Let $\mathscr{H}$ be the (abstract) completion of $\mathscr{H}_0$. Since $\mathscr{H}_0$ is a (dense) subset of $\mathscr{H}$ we can consider the mapping

$$x\to K_x$$

as a mapping from S to $\mathscr{H}$. For any element f of $\mathscr{H}$ we form the associated function on S given by

$$f(x)=(f,K_x).$$

This determines a linear mapping of $\mathscr{H}$ to a linear space of function on S with the property that elements of $\mathscr{H}_0$ are mapped into themselves. The mapping is one-to-one, for if f in $\mathscr{H}$ is mapped into 0, then f is orthogonal to every K_x, hence orthogonal to the dense subset $\mathscr{H}_0$, from which it follows that f is the 0-element of $\mathscr{H}$.

The linear space of functions obtained in this way is now identified with the completion $\mathscr{H}$ of $\mathscr{H}_0$ and this space is called the *reproducing kernel space* associated with the *reproducing kernel* or *kernel function* $K(x,y)$. Thus a positive matrix and a reproducing kernel are one and the same thing. However, it is customary to use the term positive matrix when interest is principally directed toward the function $K(x,y)$ itself, while in cases where the associated Hilbert space of functions is the object of interest the term reproducing kernel is more commonly used.

Suppose, next, that S is an arbitrary set and that $\mathscr{H}$ is a linear space of function on S which is also a Hilbert space. Let us further suppose that $\mathscr{H}$ has the property that the valuation functional is a continuous linear functional on $\mathscr{H}$, that is to say, that for every x in S the linear functional

$$L_x(f) = f(x)$$

is continuous in the element f. It follows immediately that L_x can be represented as an inner product, that is, that there exists a unique element K_x in the space such that $f(x) = (f, K_x)$. It is then easy to show that the space is a reproducing kernel space associated with the kernel function $K(x,y) = (K_y, K_x)$.

We turn now to some examples of positive matrices and reproducing kernel spaces.

Let S be an arbitrary set and $f(x)$ an arbitrary complex-valued function on it; then

$$K(x,y) = f(x)\overline{f(y)}$$

defines a very simple positive matrix on $S \times S$. The corresponding reproducing kernel space is one-dimensional and contains the function $f(x)$. More generally, since the sum of positive matrices is again a positive matrix, any finite sum of the form

$$K(x,y) = \sum_{k=1}^{N} f_k(x)\overline{f_k(y)}$$

is a positive matrix corresponding to a finite-dimensional reproducing kernel space. An infinite sum of the same form

$$K(x,y) = \sum f_k(x)\overline{f_k(y)}$$

determines a positive matrix provided the sum $K(x,x)$ is finite for all x in S. This can easily be seen from the fact that such a sum determines a mapping of S into the Hilbert space l^2, carrying x in S into the sequence $K_x = \{f_k(x)\}$ in l^2, whence $K(x,y) = (K_y, K_x)$ is necessarily a positive matrix.

Let S be the right half-axis and $L^2(S)$ the Hilbert space of functions (Lebesgue) integrable square on S. The mapping of S into $L^2(S)$ which carries x into $\chi_x(t)$, the characteristic function of the interval $(0,x)$, gives rise to the positive matrix

$$K(x,y) = (\chi_y, \chi_x) = \min(x,y)$$

which has considerable significance in probability theory. Similarly, the mapping of S into $L^2(S)$ determined by $x \to e^{-xt}$ gives rise to the positive matrix

$$K(x,y) = \frac{1}{x+y}.$$

Further examples of positive matrices will appear frequently in the sequel.

Let $\mathscr{H}$ be a reproducing kernel space associated with S and a kernel $K(x,y)$ and let $u_n(x)$ be a complete orthonormal system in $\mathscr{H}$. We consider the canonical mapping of S into $\mathscr{H}$ and for a fixed y in S we form the Fourier expansion of K_y relative to the orthonormal set. Accordingly

$$K_y = \sum (K_y, u_n) u_n$$

and the system of Fourier coefficients $(K_y, u_n) = \overline{u_n(y)}$ is square summable. It follows that the inner product (K_y, K_x) may be written

$$K(x,y) = (K_y, K_x) = \sum u_n(x) \overline{u_n(y)}$$

where the series converges absolutely. It is essential to notice that this identity does not depend on the choice of the complete orthonormal set. Hitherto we have made no hypotheses concerning the set S. Let us suppose now that S is a separable metric space. In this case, if $K(x,y)$ is continuous on $S \times S$, then every function $f(x)$ in the reproducing kernel space will be continuous on S. For given any pair of points x and y in S we have

$$|f(x) - f(y)| = |(f, K_x - K_y)| \leqq \|f\| \|K_x - K_y\|$$

and as y approaches x the quantity on the right converges to 0, since

$$\|K_x - K_y\|^2 = K(x,x) + K(y,y) - 2[\operatorname{Re} K(x,y)].$$

Thus every function $f(x)$ in the space is continuous. Because S is separable there is a countable dense subset S' of S to which corresponds a countable set of elements K_x in $\mathscr{H}$ with x in S'. The linear span of these elements is the whole reproducing kernel space, since a function which was orthogonal to every K_x with x in S' would have to vanish on the dense S'. To these functions we can apply the Gram-Schmidt process and construct a complete orthonormal system $u_n(x)$ in the space. The functions of this particular complete orthonormal set will be finite linear combinations of sections of the kernel function $K(x,y)$ with y in S'. Accordingly, such functions will inherit certain properties of the kernel. In the future we will occasionally invoke such special complete orthonormal sets in a reproducing kernel space. Of course, this will be legitimate only when S is a separable metric space and the kernel continuous on the product $S \times S$.

It is often interesting to study those reproducing kernel spaces for which S is a connected open set $\mathscr{D}$ in the complex plane and such that every function in the reproducing kernel space is analytic in $\mathscr{D}$. In such a case, since the function $K(z,w)$ for any fixed w is in the Hilbert space, $K(z,w)$ is analytic in z. However, since $K(z,w)=\overline{K(w,z)}$, $K(z,w)$ for fixed z is the conjugate of an analytic function of w. This can also be put in the following form: $K(z,w)$ is analytic in $\bar{w}$. Functions of this kind are called *sesqui-analytic*. More formally: a function $F(z,w)$ of two complex variables, defined on the product $\mathscr{G}_1 \times \mathscr{G}_2$ where $\mathscr{G}_1$ and $\mathscr{G}_2$ are open subsets of the complex plane is sesqui-analytic if it is analytic in z and also analytic in $\bar{w}$. Equivalently, $F(z,w)$ is sesqui-analytic if and only if the function $G(z,w)=F(z,\bar{w})$ defined on $\mathscr{G}_1 \times \overline{\mathscr{G}}_2$ is an analytic function of the two complex variables z and w. It should now be obvious what should be meant by the term sesqui-analytic continuation of a sesqui-analytic function $F(z,w)$: it is the continuation obtained by considering an analytic continuation of the analytic $G(z,w)$ and then passing to the function $F(z,w)=G(z,\bar{w})$.

We are led to another definition. Let $\mathscr{D}$ be an open subset of the complex plane; a function $K(z,w)$ defined on $\mathscr{D} \times \mathscr{D}$ is a *Bergman kernel function* if and only if it is a positive matrix which is also sesqui-analytic. We have already proved half of the following theorem.

Theorem. *Let $\mathscr{H}$ be a reproducing kernel space of functions defined on $\mathscr{D}$, an open subset of the complex plane. Every function in $\mathscr{H}$ is analytic if and only if the corresponding kernel function is a Bergman kernel function.*

Proof. We will suppose that the kernel is a Bergman kernel function and deduce that the space consists of analytic functions. We first set up an appropriate complete orthonormal system as follows. Let $\{w_n\}$ be a countable dense subset of $\mathscr{D}$ and apply the Gram-Schmidt process to the sequence of functions $K(z,w_n)$ to obtain a complete orthonormal system $\{u_n(z)\}$ in $\mathscr{H}$ which consists of functions analytic in $\mathscr{D}$. The most general function in the space may be written

$$f(z)=\sum a_n u_n(z)$$

where $\sum |a_n|^2=\|f\|^2$ is finite, and we must show that this $f(z)$ is analytic. The partial sums of the series are certainly analytic and for fixed z are bounded by $\|f\|\,\|K_z\|=\|f\|\sqrt{K(z,z)}$. Let w be arbitrary in $\mathscr{D}$ and let $\mathscr{C}$ be an open disk of radius r about w as center having its closure entirely in $\mathscr{D}$. If M is the finite quantity $\sup_{z\in\mathscr{C}}\sqrt{K(z,z)}$ then the partial sums of the series are uniformly bounded in absolute value on $\mathscr{C}$ by $\|f\|M$. It follows that the partial sums have a subsequence converging uniformly on a given compact subset of $\mathscr{C}$, for example, the closed disk of radius $r/2$ about w. Because the series already converges, the series, and the selected subsequence of partial sums converge to the same limit $f(z)$ in that smaller disk. Thus $f(z)$ in a neighborhood of w appears as the uniform limit of a sequence of analytic functions, and is therefore itself analytic in such a neighborhood. Because w is arbitrary, $f(z)$ is analytic throughout $\mathscr{D}$ and the proof is complete.

We have already met an important Bergman kernel associated with a function $\varphi(z)$ in the class $P(a,b)$. This was the kernel

$$\begin{aligned} K(z,w) &= \frac{\varphi(z)-\overline{\varphi(w)}}{z-\overline{w}} \\ &= \alpha + \int \frac{1}{\lambda-z}\,\overline{\frac{1}{\lambda-w}}\,d\mu(\lambda)\,. \end{aligned}$$

The corresponding open set $\mathscr{D}$ is large: it consists of the union of the upper half-plane, the interval (a,b) and the lower half-plane.

Lemma. *Let $K(x,y)$ and $H(x,y)$ be positive matrices defined on $S\times S$; then their product*

$$G(x,y)=K(x,y)H(x,y)$$

is also a positive matrix on $S\times S$.

Proof. It is enough to show that $G(x,y)$ is a positive matrix over $F\times F$ where F is an arbitrary finite subset of S. But this is an immediate consequence of the theorem in Chapter I that the Schur product of positive matrices is again a positive matrix. Nevertheless, it is interesting to give another proof, independent of Schur's Theorem, and implying it. We know that there exist convergent representations for $K(x,y)$ and $H(x,y)$, namely

$$K(x,y)=\sum u_n(x)\overline{u_n(y)} \quad \text{and} \quad H(x,y)=\sum v_m(x)\overline{v_m(y)}$$

and therefore

$$G(x,y)=\sum\sum(u_n(x)v_m(x))(\overline{u_n(y)v_m(y)})$$

is evidently a positive matrix if the series converges for $x=y$. It is obvious that in this case we obtain a series of positive numbers converging to the product $K(x,x)H(x,x)$.

As a consequence of Schur's Theorem we infer, of course, that if $K(x,y)$ is a positive matrix, then so is

$$E(x,y)=\exp K(x,y)\,.$$

Chapter XI. Nagy-Koranyi Proof of Loewner's Theorem

In this chapter we apply the elementary theory of reproducing kernels developed in the previous chapter to establish a theorem due to G. Pick which complements the assertions of a theorem of Chapter III. We also give another proof of Loewner's theorem. Both of these arguments are essentially due to Koranyi and Nagy.

Pick's Theorem. *Let S be a finite set of points in the upper half-plane*

$$z_1, z_2, \ldots, z_n$$

and

$$w_1, w_2, \ldots, w_n$$

by equally many (not necessarily distinct) complex numbers. If

$$K_{ij} = K(z_i, z_j) = \frac{w_i - \overline{w}_j}{z_i - \overline{z}_j}$$

is a positive matrix, then there exists a rational function $\varphi(z)$ in the Pick class of degree at most n, real on the real axis and satisfying the n equations $\varphi(z_i) = w_i$.

Proof. We will first suppose that the determinant $\det K$ is not zero. We deduce that in the reproducing kernel space associated with $K(z_i, z_j)$ the elements K_{z_i} are linearly independent and hence that the space has dimension n. Since the space of all functions on S also has dimension n, those spaces must coincide. It follows that the space contains the function $f(z)$ defined by the equations $f(z_i) = w_i$. Since the K_{z_i} form a basis for the space we can define a linear transformation H of the space into itself by setting

$$H K_{z_i} = \overline{z}_i K_{z_i} + f.$$

It is easy to verify that $H = H^*$, since if $u = \sum a_i K_{z_i}$ then

$$(Hu, u) = \sum \sum a_i \overline{a}_j (\overline{z}_i K_{z_i} + f, K_{z_j})$$

and we have only to show that the matrix element is Hermitian symmetric.

That is to say, we must show the equality

$$\bar{z}_i(K_{z_i}, K_{z_j}) + f(z_j) = z_j\overline{(K_{z_j}, K_{z_i})} + \overline{f(z_i)}$$

and this easily reduces to

$$(z_i - \bar{z}_j)\, K(z_i, z_j) = w_i - \bar{w}_j$$

which is obvious.

We may therefore write

$$(H - \bar{z}_i I)\, K_{z_i} = f$$

and deduce that $R_{\bar{z}_i} f = K_{z_i}$ where R_z is the resolvent of H. Accordingly

$$(R_{\bar{z}_i} f, f) = (K_{z_i}, f) = \overline{f(z_i)}$$

and therefore $\varphi(z) = (R_z f, f)$ is a rational function of degree at most n which is real on the real axis and which satisfies the equations $\varphi(\bar{z}_i) = \bar{w}_i$. Hence in view of the reflection principle, $\varphi(z_i) = w_i$ for all i and the theorem has been proved when $\det K$ is not 0.

When $\det K = 0$ we alter the problem slightly, substituting $w_i + \varepsilon \log z_i$ for w_i where ε is small and positive and the obvious determination of logarithm taken. The Pick matrix associated with the new interpolation problem is of the form

$$K'' = K + \varepsilon K'$$

where K' is the Pick matrix associated with the logarithm function and the set S, hence a positive definite matrix. It follows that K'' is a positive matrix with a non-vanishing determinant, and so there exists a rational $\varphi_\varepsilon(z)$ real on the real axis with $\varphi_\varepsilon(z_i) = w_i + \varepsilon \log z_i$. As ε approaches 0 the system of Pick functions $\varphi_\varepsilon(z)$ is uniformly bounded on the set S and in view of a theorem of Chapter II has a subsequence converging uniformly on compact subsets of the half-plane to a limiting Pick function $\varphi(z)$. Clearly $\varphi(z_i) = w_i$ for every i, and the theorem of Chapter III guarantees that this function is rational, of degree lower than n and that it is real on the real axis. This completes the proof of the theorem.

When the set S is infinite there is a corresponding, but weaker theorem.

Theorem I. *Let S be a subset of the upper half-plane and $f(z)$ a complex valued function defined on S. Let the function*

$$K(z, w) = \frac{f(z) - \overline{f(w)}}{z - \bar{w}}$$

be a positive matrix on $S \times S$. Then there exists a Pick function $\varphi(z)$ which coincides with $f(z)$ on S.

Proof. We first prove the theorem in the easy case that S is countable. We may suppose the points of S enumerated, and for each n we invoke the previous theorem to find a suitable $\varphi_n(z)$ in the Pick class satisfying the n equations $\varphi_n(z_i)=f(z_i)$. This sequence of Pick functions is uniformly bounded at the point z_1 and therefore has a convergent subsequence which of course converges to a Pick function $\varphi(z)$ which coincides with $f(z)$ at the points of the sequence S.

When the set S is not countable we select a countable dense subset S' and find a Pick function coinciding with $f(z)$ on S'. This function is continuous on S, and we will presently show that $f(z)$ is also continuous on S. Thus $f(z)-\varphi(z)$ is continuous on S and vanishes on S' and $\varphi(z)$ satisfies the conditions of the theorem.

We know that $f(z)$ is continuous on S because if it were not, there would exist a z_0 in S and a sequence $z_1, z_2, \ldots$ converging to z_0 such that $f(z_n)$ does not converge to $f(z_0)$. This, however, is impossible, since our theorem has been proved for countable sets, and there therefore exists a Pick function coinciding with f on the sequence z_n as well as the point z_0. The Pick function is continuous, and hence $f(z)$ is too. This completes the proof of the theorem.

Essentially the same technique is used in the following proof of Loewner's Theorem which depends on the theory of reproducing kernels and some facts about the calculus of self-adjoint operators on a separable Hilbert space. We state the theorem again, but in a trivially different form.

Theorem II. *Let $f(x)$ be a C^1-function defined on the interval (a,b) such that the function*

$$\begin{aligned} K(x,y) &= \frac{f(x)-f(y)}{x-y} \qquad \text{for } x \neq y \\ &= f'(x) \qquad\qquad \text{for } x=y \end{aligned}$$

is a positive matrix on the product $(a,b)\times(a,b)$. Then $f(x)$ belongs to the class $P(a,b)$.

Proof. There are certain preliminary simplications we can make. We may suppose that $f'(x)$ is strictly positive since if $f'(x_0)=0$ then the inequality

$$K(x,x_0)^2 \leqq K(x,x)\,K(x_0,x_0)$$

implies that the function $f(x)$ is a constant, hence evidently in $P(a,b)$. Moreover, if $\varphi(x)$ is a function in some Pick class $P(c,d)$ so that the composed function $g(x)=(\varphi\circ f)(x)$ makes sense and is continuous on (a,b) then $g(x)$ is also associated with a positive matrix $G(x,y)$ since we may write

$$G(x,y) = \frac{g(x)-g(y)}{x-y} = \frac{\varphi(f(x))-\varphi(f(y))}{f(x)-f(y)} \cdot \frac{f(x)-f(y)}{x-y}$$

and

$$G(x,x) = g'(x) = \varphi'(f(x))\, f'(x)$$

and invoke the fact that the Schur product of positive matrices is again a positive matrix. In a similar way, if the function $\varphi(x)$ is in $P(c,d)$ and takes values in (a,b) we show that $(f\circ\varphi)(x)$ is in $P(c,d)$. There is therefore no loss of generality in supposing that the interval (a,b) is the interval $(-1,1)$. Finally, we are at liberty to take the function as vanishing at the origin with derivative $+1$ there. It follows that the function to be investigated may be taken to be a function $g(x)$ on $(-1,1)$ satisfying the conditions $g(0)=0$ and $g'(0)=1$ and this is what is usually done in proofs of the Loewner theorem. However, we shall proceed a step farther and compose this $g(x)$ with a linear fractional transformation in the Pick class to obtain

$$f(x) = g(-1/x)$$

which by hypothesis gives rise to a positive matrix $K(x,y)$ defined on $S\times S$ where S is the set of all real x such that $|x|>1$. We note that $f(x)$ vanishes at infinity and is negative for $x>1$ while it is positive for $x<-1$.

Let $\mathscr{H}$ be the reproducing kernel space associated with $K(x,y)$. In that space we consider the system $-x\,K_x$ as x in S approaches infinity. We first show that this system approaches a limit. Indeed the system is a Cauchy system: if $|x|$ and $|y|$ are sufficiently large the norm of the difference $-x\,K_x + y\,K_y$ is arbitrarily small. We have

$$\begin{aligned}\|-x\,K_x + y\,K_y\|^2 &= x^2 K(x,x) + y^2 K(y,y) - 2xy\,K(x,y)\\ &= x^2 f'(x) + y^2 f'(y) - 2xy\,\frac{f(x)-f(y)}{x-y}.\end{aligned}$$

Since $f'(z) = g'(-1/z)/z^2$ this becomes

$$g'(-1/x) + g'(-1/y) - 2\,\frac{g(-1/x)-g(-1/y)}{-1/x+1/y}$$

and according to the Mean Value theorem this is equal to

$$g'(-1/x) + g'(-1/y) - 2g'(\theta)$$

where θ is some value in between $-1/x$ and $-1/y$. Because $g'(z)$ is continuous near the origin, this quantity converges to 0 with $1/x$ and $1/y$.

Let f^* be the element of $\mathscr{H}$ to which $-x\,K_x$ converges as $|x|$ increases without bound. This limit is an element of a reproducing kernel space, and so is a function on S. We determine that function next.

$$f^*(y)=(f^*,K_y)=\lim_{|x|\to\infty}(-xK_x,K_y)=\lim_{|x|\to\infty}-x\frac{f(x)-f(y)}{x-y}=f(y).$$

Hence the function f of the theorem is itself an element of the reproducing kernel space and is the limit of $-xK_x$ as $|x|$ approaches infinity.

Let h be an arbitrary element of the reproducing kernel space. Evidently

$$(h,f)=\lim_x(h,-xK_x)=-\lim xh(x)$$

and hence the functions $h(x)$ in the reproducing kernel space have the property that this limit exists: $\lim xh(x)$ exists as $|x|$ approaches infinity and is some real number, generally not 0.

This circumstance leads us to introduce a linear transformation T from $\mathscr{H}$ to the space of all continuous functions on S, vanishing at infinity. The transformation is defined by the equation

$$(Th)(x)=xh(x)+(h,f).$$

The linearity of T is evident. We next show that T has a very good behavior on the dense subspace $\mathscr{H}_0$ of $\mathscr{H}$ determined by the kernel function itself. $\mathscr{H}_0$ is the space of finite linear combinations of elements K_y and it is easy to verify that $TK_y=yK_y+f$. Hence T carries $\mathscr{H}_0$ into $\mathscr{H}$.

Next we consider H_0, the restriction of T to $\mathscr{H}_0$. This is a symmetric operator, since, as we shall immediately see, the numbers (H_0u,u) are real for every u in $\mathscr{H}_0$. Indeed, if $u=\Sigma a_k K_{y_k}$ then

$$(H_0u,u)=\sum\sum a_k\bar{a}_j[y_kK(y_j,y_k)+f(y_j)]$$

and the quadratic form corresponds to a real symmetric matrix, since the real quantities $y_kK(y_j,y_k)+f(y_j)$ and $y_jK(y_k,y_j)+f(y_k)$ are equal.

It is easy to determine the effect of the adjoint H_0^* on an element $h(x)$ in its domain. We will have

$$(H_0^*h)(x)=(H_0^*h,K_x)=(h,H_0K_x)=(h,xK_x+f)=xh(x)+(h,f).$$

Thus H_0^* acts on $h(x)$ in the same way that T does.

Let $u(x)$ be an element of $\mathscr{H}$ for which $H_0^*u=iu$. We then have

$$(x-i)u(x)=-(u,f)=c$$

and therefore

$$u(x)=\frac{c}{x-i}.$$

Of course, it may be that no such $u(x)$ exists in the space $\mathscr{H}$. In any event, it is clear that the multiplicity of i as an eigenvalue of H_0^* is either 0 or 1. If it is 1, the function $v(x)=1/(x+i)$ is in the reproducing kernel space and occurs as an eigenfunction of H_0^* with eigenvalue $-i$. It follows that the deficiency indices of H_0 are either $(0,0)$ or $(1,1)$ and in particular, that those deficiency indices are equal. Hence, H_0 possesses at least one self-adjoint extension. We select such a one and denote it with the letter H.

Let y be a real number for which $|y|>1$. From $HK_y = yK_y + f$ we deduce $(H-yI)K_y = f$ where I is the identity operator. Now the operator $(H-yI)$ has no null space, since $(H-yI)v=0$ would imply that $v(x)$ was of the form $v(x)=c/(x-y)$, a function that is not continuous on S if c is not 0. It follows that the inverse operator $(H-yI)^{-1}$ exists and is self-adjoint, although it may not be a bounded operator. If E_λ is the resolution of the identity corresponding to H then

$$(H-yI)^{-1} = \int \frac{1}{\lambda - y}\, dE_\lambda$$

and the assertion that f is in the domain of that operator is equivalent to the finiteness of the integral

$$\int \frac{d(E_\lambda f, f)}{(\lambda - y)^2} = \|(H-yI)^{-1} f\|^2 = \|K_y\|^2 = K(y,y)\,.$$

Let $\mu(\lambda) = (E_\lambda f, f)$. From the finiteness of this integral we deduce that the quantity

$$\int_y^{y+\varepsilon} \frac{d\mu(\lambda)}{(\lambda - y)^2}$$

converges to 0 as the positive ε diminishes. Obviously, however, this quantity is at least as large as

$$\frac{1}{\varepsilon}\,\frac{\mu(y+\varepsilon)-\mu(y)}{\varepsilon}$$

and we infer that the right hand derivative of $\mu(\lambda)$ exists and is 0 for $\lambda = y$. A similar argument shows that the left hand derivative also vanishes for $|y|>1$. It follows that $\mu(\lambda)$ is constant on the intervals $y>1$ and $y<-1$, hence that the measure $d\mu$ is concentrated on the closed interval $[-1,1]$.

Since $(H-yI)^{-1} f = K_y$

$$((H-yI)^{-1} f, f) = (K_y, f) = \overline{f(y)} = f(y) = \int_{-1}^{1} \frac{d\mu(\lambda)}{\lambda - y}\,.$$

This formula shows that $f(x)$ is a Pick function corresponding to a measure supported by the interval $[-1,1]$. The proof is complete.

Chapter XII. The Cauchy Interpolation Problem

In this chapter we study an interpolation problem of the following type. A finite set of interpolating points is given, and the first several terms of the Taylor expansion of a rational function about each interpolation point are prescribed. As we have seen in Section 2 of Chapter I, there exists a polynomial solution to this interpolation problem. Here, however, we shall require that the degree of the rational function be sufficiently low for it to be admitted as a solution to the interpolation problem. We can state our problem more formally as follows:

Let there be given l interpolation points in the complex plane

$$z_1, z_2, \ldots, z_l$$

and equally many nonnegative integers

$$\nu_1, \nu_2, \ldots, \nu_l$$

and N numbers f_{ij} where $1 \leqq i \leqq l$ and $0 \leqq j \leqq \nu_i$ with $N = \sum_{i=1}^{l} (\nu_i + 1)$. It is required to find a rational function $f(z)$ of degree no greater than $N/2$ for which the N equations

$$f^{(j)}(z_i) = f_{ij}$$

are satisfied.

Let us first note the following important circumstance. If there exists a solution to the interpolation problem with degree smaller than $N/2$ then that solution is the unique solution to the problem. For if there were another solution $g(z)$, the difference $f(z) - g(z)$ would be rational, of degree smaller than N and having zeros of total order at least N. Thus the difference would vanish identically. When the integer N is odd, this is of course always the case, since $N/2$ is then not an integer. It follows that the interpolation problem is essentially different in the case N even and the case N odd. In most of this chapter we shall consider only the case when N is even, and we shall always take $N = 2n$.

Suppose $f(z)$ is a solution to the interpolation problem. This function may be written $f(z) = \sigma(z)/\tau(z)$ where the polynomials $\sigma(z)$ and $\tau(z)$

are relatively prime and of degree at most n. We therefore differentiate the product

$$\sigma(z) = f(z)\,\tau(z)$$

to obtain the $2n$ linear equations relating the polynomials $\sigma(z)$ and $\tau(z)$ which follow.

$$\sigma^{(k)}(z_i) = \sum_{j=0}^{k} \frac{k!}{j!(k-j)!} f_{ij}\,\tau^{(k-j)}(z_i) \qquad 0 \leqq k \leqq \nu_i\,.$$

Accordingly there is associated with our interpolation problem a problem in linear algebra which we call the corresponding linear problem: to find a pair of polynomials $[\sigma;\tau]$ each of degree at most n satisfying the $2n$ equations above. Clearly every solution to the interpolation problem gives rise to a solution of the linear problem, the converse, however, need not be true, since a solution $[\sigma;\tau]$ of the linear problem may be such that both polynomials possess a common zero at one of the interpolation points. Their ratio, then, may not be a rational function satisfying the conditions of the interpolation problem. However, it is easy to show that if the polynomial $\tau(z)$ vanishes at none of the interpolation points then the ratio $f(z)=\sigma(z)/\tau(z)$ is in fact a solution to the interpolation problem. From the equation

$$\sigma(z_i) = f_{i0}\,\tau(z_i)$$

and the fact that $\tau(z_i)$ is not zero we deduce that

$$f_{i0} = \sigma(z_i)/\tau(z_i) = f(z_i)$$

and then argue inductively, differentiating $\sigma(z)=f(z)\tau(z)$ to show that $f^{(k)}(z_i)=f_{ik}$ for all $k \leqq \nu_i$. We will have

$$\sigma^{(k)}(z_i) = f_{ik}\,\tau(z_i) + \sum_{j=0}^{k-1} \frac{k!}{j!(k-j)!} f_{ij}\,\tau^{(k-j)}(z_i)$$

as well as

$$\sigma^{(k)}(z_i) = f^{(k)}(z_i)\,\tau(z_i) + \sum_{j=0}^{k-1} \frac{k!}{j!(k-j)!} f^{(j)}(z_i)\,\tau^{(k-j)}(z_i)\,.$$

From the inductive hypothesis it follows that the sums appearing on the right sides of these equations are equal, and we deduce that

$$f_{ik}\,\tau(z_i) = f^{(k)}(z_i)\,\tau(z_i)$$

and since $\tau(z_i)$ is not 0 the result follows.

Let $\mathcal{M}$ be the vector space consisting of polynomial pairs $[\sigma;\tau]$ where the degree of the polynomials in the pair is at most n. This space evidently has dimension $2n+2$. The equations of the linear problem then distinguish a subspace $\mathcal{S}$ of $\mathcal{M}$ consisting of the pairs which satisfy those equations. We call $\mathscr{S}$ the space of solutions to the linear problem and note that it has dimension at least 2, since there are only $2n$ conditions in the linear problem. Accordingly, $\mathscr{S}$ is not empty, and the linear problem always possesses solutions.

We shall say that the linear problem has a solution of lower degree if there exists a non-trivial pair $[\sigma;\tau]$ in $\mathscr{S}$ where the degree of both $\sigma(z)$ and $\tau(z)$ is smaller than n. In a similar way we shall say that the interpolation problem has a solution of lower degree if it has a rational solution $f(z)$ with degree f smaller than n. We have already observed that if a solution of lower degree exists it is the unique solution to the interpolation problem.

Lemma 1. *Let $[\sigma;\tau]$ be a non-trivial solution to the linear problem where the polynomials $\sigma(z)$ and $\tau(z)$ have a common zero at $z=\lambda$. Put*

$$\sigma(z)=(z-\lambda)\hat{\sigma}(z)$$

and

$$\tau(z)=(z-\lambda)\hat{\tau}(z).$$

Then the polynomials $\hat{\sigma}(z)$ and $\hat{\tau}(z)$ are also solutions to the linear problem if λ is not one of the interpolation points, while if λ is an interpolation point those polynomials satisfy all of the equations of the linear problem except perhaps the equation at λ involving the derivative of highest order.

Proof. We have

$$\sigma^{(k)}(z)=(z-\lambda)\hat{\sigma}^{(k)}(z)+k\hat{\sigma}^{(k-1)}(z)$$

and a similar equation for $\tau(z)$. The equations satisfied by $\sigma(z)$ and $\tau(z)$ may therefore be put in the form

$$(z_i-\lambda)\left\{\hat{\sigma}^{(k)}(z_i)-\sum_{j=0}^{k}\frac{k!}{j!(k-j)!}f_{ij}\hat{\tau}^{(k-j)}(z_i)\right\}$$

$$=-k\left\{\hat{\sigma}^{(k-1)}(z_i)-\sum_{j=0}^{k-1}\frac{(k-1)!}{j!(k-1-j)!}f_{ij}\hat{\tau}^{(k-1-j)}(z_i)\right\}$$

If λ is equal to no z_i we put $k=0$ to obtain the lowest order equation at z_i for the pair $[\hat{\sigma};\hat{\tau}]$. Next we put $k=1$ to obtain in the next equation at that point, and successively all of the v_i+1 equations at z_i for the pair $[\hat{\sigma};\hat{\tau}]$. On the other hand, if $\lambda=z_i$ for some i, the left side of the equation above vanishes for all admissible values of k and we obtain v_i equations which the pair $[\hat{\sigma};\hat{\tau}]$ satisfies. It is not clear whether or not that pair satisfies the remaining equation.

The equations used in the proof of the previous lemma make it clear that if the pair $[\sigma;\tau]$ satisfy the equations of the linear problem so also do the polynomials $(z-\lambda)\sigma(z)$ and $(z-\lambda)\tau(z)$ although these new polynomials may have degree greater than n. Since λ is arbitrary, we deduce that for any polynomial $h(z)$ the polynomials $h(z)\sigma(z)$ and $h(z)\tau(z)$ also satisfy the equations of the linear problem. We will use this simple remark in the sequel.

Lemma 2. *Let* $[\sigma_0;\tau_0]$ *and* $[\sigma_1;\tau_1]$ *be solutions to the linear problem and form the determinant*

$$\delta(z)=\begin{vmatrix}\sigma_0(z) & \sigma_1(z)\\ \tau_0(z) & \tau_1(z)\end{vmatrix}.$$

Then $\delta(z)=C\Delta(z)$ *for some constant* C *where*

$$\Delta(z)=\prod_{i=0}^{l}(z-z_i)^{\nu_i+1}$$

Proof. Since the degree of $\delta(z)$ is at most $2n=N$ it is enough for us to show that at each interpolation point z_i the equation $\delta^{(k)}(z_i)=0$ is valid for $0\leqq k\leqq \nu_i$. We may suppose that $z_i=z_1$ and compute as follows.

$$\delta^{(k)}(z_1)=\sum_{i=0}^{k}\frac{k!}{i!(k-i)!}\left[\sigma_0^{(i)}(z_1)\tau_1^{(k-i)}(z_1)-\sigma_1^{(i)}(z_1)\tau_0^{(k-i)}(z_1)\right].$$

Since

$$\sigma_0^{(i)}(z_1)=\sum_{j=0}^{i}\frac{i!}{j!(i-j)!}f_{1j}\tau_0^{(i-j)}(z_1)$$

and a similar equation holds for $\sigma_1^{(i)}(z_1)$ we have $\delta^{(k)}(z_1)$ equal to

$$\sum_{i=0}^{k}\sum_{j=0}^{i}\frac{k!\,i!}{i!(k-i)!\,j!(i-j)!}f_{1j}\left[\tau_0^{(i-j)}(z_1)\tau_1^{(k-i)}(z_1)-\tau_1^{(i-j)}(z_1)\tau_0^{(k-i)}(z_1)\right].$$

The coefficient of f_{1j} in this expression is

$$\frac{k!}{j!}\sum_{i=j}^{k}\left[\frac{\tau_0^{(i-j)}(z_1)\tau_1^{(k-i)}(z_1)}{(i-j)!\,(k-i)!}-\frac{\tau_1^{(i-j)}(z_1)\tau_0^{(k-i)}(z_1)}{(i-j)!\,(k-i)!}\right]$$

and this vanishes identically.

The previous lemma makes possible a complete description of the state of affairs when the linear problem has a solution of lower degree. Suppose that $[\sigma_0;\tau_0]$ is a non-trivial solution of lower degree, both polynomials in the pair then having degree at most $n-1$. We select an arbitrary non-trivial solution $[\sigma_1;\tau_1]$ and form the determinant

$$\delta(z)=\begin{vmatrix}\sigma_0(z) & \sigma_1(z)\\ \tau_0(z) & \tau_1(z)\end{vmatrix}.$$

This polynomial vanishes identically in view of the previous lemma and the fact that its degree is at most $2n-1$. It follows that the ratios

$$\sigma_0(z)/\tau_0(z)\quad\text{and}\quad\sigma_1(z)/\tau_1(z)$$

are equal for all but finitely many values of z, the exceptional values being those where the denominators vanish. Both solutions to the linear problem therefore determine the same rational function $f(z)$ of degree at most $n-1$. Since the pair $[\sigma_1; \tau_1]$ was arbitrary in $\mathscr{S}$ it follows that all non-trivial solutions to the linear problem give rise to the same rational function $f(z)$. We may write $f(z)$ as the ratio of two relatively prime polynomials of degree at most $n-1$, putting $f(z)=\hat{\sigma}(z)/\hat{\tau}(z)$ although we do not know that the pair $[\hat{\sigma}; \hat{\tau}]$ is a solution to the linear problem. Every solution to the linear problem is then of the form $h(z)\hat{\sigma}(z)$ and $h(z)\hat{\tau}(z)$ where the polynomial $h(z)$ is of sufficiently low degree. The collection of polynomials $h(z)$ so obtained form a linear space $\mathscr{H}$ and we can find a non-trivial polynomial $h_0(z)$ in that space of minimal degree. If $h_0(z)$ is a constant the pair $[\hat{\sigma}; \hat{\tau}]$ is a solution to the linear problem, but if $h_0(z)$ is of higher degree this will not be the case. It is easy to see that $h_0(z)$ divides every other polynomial in $\mathscr{H}$ since for any such polynomial $h(z)$ we may write

$$h(z)=q(z)h_0(z)+r(z)$$

where the degree of $r(z)$ is smaller than that of $h_0(z)$. Since the degree of $q(z)$ is not too large, the remarks following Lemma 1 make it clear that the product $q(z)h_0(z)$ also belongs to $\mathscr{H}$ and hence that $r(z)$ does too. It follows that $r(z)$ is the zero polynomial, since h_0 was of minimal degree.

It is now clear that when the linear problem admits a solution of lower degree the corresponding interpolation problem has a solution if and only if the linear problem has a solution $[\hat{\sigma}; \hat{\tau}]$ where the polynomials $\hat{\sigma}$ and $\hat{\tau}$ are relatively prime.

We next consider the case when the linear problem admits no solution of lower degree. Let L_1 be the linear functional on $\mathscr{S}$ which assigns to the pair $[\sigma; \tau]$ the coefficient of z^n in the polynomial $\sigma(z)$, and similarly let L_2 assign the corresponding coefficient for $\tau(z)$. The intersection of the null spaces of these two linear functionals is not empty if the dimension of $\mathscr{S}$ exceeds 2. Since there would then exist a solution of lower degree it follows that the dimension of $\mathscr{S}$ is exactly 2, because we have already observed that its dimension is at least 2.

If $[\sigma; \tau]$ is a non-trivial element of $\mathscr{S}$ then at least one of the polynomials $\sigma(z)$ and $\tau(z)$ has degree n. Similarly if $[\sigma_0; \tau_0]$ and $[\sigma_\infty; \tau_\infty]$ are two linearly independent elements of the space, at least one of the polynomials $\sigma_0(z)$ and $\sigma_\infty(z)$ has degree n, because otherwise all polynomials $\sigma(z)$ would have degree at most $n-1$, and at least one of the non-trivial $\tau(z)$ would have a lower degree as well, thus giving rise to a solution of lower degree. In a similar way we see that at least one of the polynomials $\tau_0(z)$ and $\tau_\infty(z)$ must have degree n.

These considerations permit us to show that if $[\sigma_0; \tau_0]$ and $[\sigma_\infty; \tau_\infty]$ form a basis for $\mathscr{S}$ then the corresponding determinant

$$\delta(z) = \begin{vmatrix} \sigma_0(z) & \sigma_\infty(z) \\ \tau_0(z) & \tau_\infty(z) \end{vmatrix}$$

cannot vanish identically. To show this we note first, arguing from the remarks made above, that if one of the four polynomials in the determinant has degree smaller than n, then the coefficient of z^{2n} in $\delta(z)$ is not 0. On the other hand, if all four polynomials have degree n, we may write

$$\delta(z) = \begin{vmatrix} \sigma_0(z) + t\sigma_\infty(z) & \sigma_\infty(z) \\ \tau_0(z) + t\tau_\infty(z) & \tau_\infty(z) \end{vmatrix}$$

and for some value of t either

$$\begin{aligned} \sigma_t(z) &= \sigma_0(z) + t\sigma_\infty(z) && \text{or the corresponding} \\ \tau_t(z) &= \tau_0(z) + t\tau_\infty(z) && \text{has lower degree.} \end{aligned}$$

Thus, again, the leading coefficient in $\delta(z)$ is not zero.

Lemma 3. *Suppose that the linear problem has no solution of lower degree and let $[\sigma_0; \tau_0]$ and $[\sigma_\infty; \tau_\infty]$ be a basis for the two-dimensional space of solutions. Put*

$$\sigma_t(z) = \sigma_0(z) + t\sigma_\infty(z)$$

and

$$\tau_t(z) = \tau_0(z) + t\tau_\infty(z).$$

Then for no two distinct values of t do the polynomials $\tau_t(z)$ have a common zero.

Proof. It is easy to see that if two of the polynomials $\tau_t(z)$ have a common zero at $z = \lambda$ then all those polynomials have a zero there. It follows that λ is a zero of the non-trivial $\delta(z)$ and therefore that λ is an interpolation point. From this it follows that $\sigma_t(\lambda) = 0$ for all values of t and, although we do not use this fact, that λ must be a zero of higher order of $\delta(z)$. We then write

$$\sigma_t(z) = (z - \lambda)\hat{\sigma}_t(z)$$

and

$$\tau_t(z) = (z - \lambda)\hat{\tau}_t(z)$$

and infer from Lemma 1 that the pair $[\hat{\sigma}_t; \hat{\tau}_t]$ satisfies all of the equations of the linear problem with the possible exception of one equation, namely the equation involving the derivative of highest order at $z = \lambda$. We shall show that for an appropriate choice of the parameter t this highest order equation is also satisfied. This will imply that the linear problem has a solution of lower degree, contradicting the hypothesis of the lemma. We may take $\lambda = z_1$ and $k = v_1$; the equation which we would like to have satisfied is

$$\hat{\sigma}_0^{(k)}(z_1) + t\hat{\sigma}_\infty^{(k)}(z_1) = \sum_{j=0}^{k} \frac{k!}{j!(k-j)!} f_{1j} [\hat{\tau}_0^{(k-j)}(z_1) + t\hat{\tau}_\infty^{(k-j)}(z_1)]$$

which may also be written

$$\hat{\sigma}_0^{(k)}(z_1) - \sum_{j=0}^{k} \frac{k!}{j!(k-j)!} f_{1j} \hat{\tau}_0^{(k-j)}(z_1)$$
$$= -t\left[\hat{\sigma}_\infty^{(k)}(z_1) - \sum_{j=0}^{k} \frac{k!}{j!(k-j)!} f_{1j} \hat{\tau}_\infty^{(k-j)}(z_1)\right].$$

If the coefficient of t vanishes, the pair $[\hat{\sigma}_\infty; \hat{\tau}_\infty]$ is a solution to the linear problem of lower degree, while if that coefficient does not vanish the equation above determines a value of t such that the corresponding $[\hat{\sigma}_t; \hat{\tau}_t]$ satisfies all of the equations of the linear problem. Thus, in any event, there is a solution of lower degree.

Accordingly, when there are no solutions of lower degree, we select a basis $[\sigma_0; \tau_0]$ and $[\sigma_\infty; \tau_\infty]$ in the solution space and construct the family $[\sigma_t; \tau_t]$. The polynomial

$$\tau_t(z) = \tau_0(z) + t\tau_\infty(z)$$

vanishes at z if and only if t is equal to $T(z) = -\tau_0(z)/\tau_\infty(z)$. Let the values which the rational function $T(z)$ assumes on the set of interpolation points be called exceptional values of the parameter. There will be at most l of these numbers, $t_i = T(z_i)$, and there may well be fewer, since the t_i need not be distinct. For any t which is not exceptional, then, the polynomial $\tau_t(z)$ is never 0 on the set of interpolation points. The polynomials $\tau_t(z)$ and $\sigma_t(z)$ are then relatively prime, since a common zero would have to be a zero of the non-trivial $\delta(z)$ and therefore an interpolation point. It follows that the ratio

$$f_t(z) = \frac{\sigma_t(z)}{\tau_t(z)} = \frac{\sigma_0(z) + t\sigma_\infty(z)}{\tau_0(z) + t\tau_\infty(z)}$$

is a rational function of degree exactly n which is a solution to the interpolation problem, and every solution to the interpolation problem is obtained in this way.

We next suppose that t is an exceptional value of the parameter, say $t = t_1 = T(z_1)$. The polynomial $\tau_t(z)$ then has a zero at z_1 and so does the corresponding $\sigma_t(z)$ since $\sigma(z_1) = f_{10}\,\tau(z_1)$ is one of the equations of the linear problem. By $f_t(z)$ in this case we must understand the ratio $\sigma_t(z)/\tau_t(z)$ when the common zeros have been cancelled out. This function, called an exceptional solution to the interpolation problem, is in fact not a solution to that problem since it is of degree smaller than n, and there are no solutions to that problem of degree smaller than n because the corresponding linear problem admits no solution of lower degree. However, the exceptional solutions enter quite naturally in the family of functions $f_t(z)$ and turn out to be very useful in the analysis of the interpolation problems studied in the following chapter.

If a point z is fixed, and is not an interpolation point, then evidently $f_t(z)$ is a linear fractional function in the variable t; it is not a constant, since the corresponding determinant $\delta(z)$ is not 0. Moreover, if z is an interpolation point, say $z=z_1$, the derivative $f_t^{(\nu_1+1)}(z_1)$ is also a nontrivial linear fractional function of t. To show this we write

$$\sigma_t(z)=f_t(z)\tau_t(z)$$

and differentiate ν_1+1 times to obtain

$$\sigma_t^{(\nu_1+1)}(z_1)=\sum_{j=0}^{\nu_1}\frac{(\nu_1+1)!}{j!(\nu_1+1-j)!}f_{1j}\tau_t^{(\nu_1+1-j)}(z_1)+f_t^{(\nu_1+1)}(z_1)\tau_t(z_1).$$

Hence if t is not the exceptional value $t_1=T(z_1)$ we have

$$f_t^{(\nu_1+1)}(z_1)=\frac{A(z_1)+tB(z_1)}{\tau_t(z_1)}$$

a linear fractional function in t. We should note that this derivative exists and is finite whenever t is not the exceptional value $t_1=T(z_1)$ associated with z_1. Moreover, the derivative is not a constant in t, since if it were, for an appropriate choice of C we would adjoin the equation

$$f^{(\nu_1+1)}(z_1)=C$$

to the original interpolation problem to obtain a new one with $2n+1$ conditions having all but finitely many of the $f_t(z)$ as solutions. This contradicts the known uniqueness of solutions to the Cauchy interpolation problem when the number N of conditions is odd.

These considerations make it clear that not only have we a great many solutions to the interpolation problem when there are no solutions of lower degree, but that we are almost entitled to add a further linear condition to the problem. Thus, given an interpolation problem with $N=2n$ conditions where the corresponding linear problem has no solution of lower degree, if we adjoin a further interpolation point z_{l+1} and prescribe the value $f(z_{l+1})$, then there exists in the family $f_t(z)$ precisely one function taking the prescribed value at the new point. However, it may happen that this $f_t(z)$ is an exceptional solution, and therefore that it fails to satisfy one or more of the conditions of the original interpolation problem. Similarly, instead of adjoining a new interpolation point, we could adjoin a further condition at one of the interpolation points and require that the derivative $f^{(\nu_i+1)}(z_i)$ take an assigned value. Here again, there will be a unique function $f_t(z)$ in the family satisfying that condition, but there is no guarantee that this is not an exceptional solution.

Let z_0 be a point in the complex plane which is not an interpolation point. There will then be exactly one value of t for which $\tau_t(z_0)=0$, namely $t=T(z_0)$. Since $\delta(z_0)$ is not zero, the corresponding $\sigma_t(z_0)$ is not zero and $f_t(z)$ has a pole at $z=z_0$. We now compute the residue of that function at z_0. Clearly this is

$$r(z_0)=\lim_{z\to z_0}(z-z_0)f_t(z).$$

We write $\tau_t(z)=(z-z_0)\tau_t'(z_0)+(z-z_0)^2H(z)$ to get

$$r(z_0)=\lim_{z\to z_0}\frac{(z-z_0)\sigma_t(z)}{(z-z_0)\tau_t'(z_0)+(z-z_0)^2H(z)}=\frac{\sigma_0(z_0)+t\sigma_\infty(z_0)}{\tau_0'(z_0)+t\tau_\infty'(z_0)}$$

and putting $t=-\tau_0(z_0)/\tau_\infty(z_0)=T(z_0)$ we finally obtain

$$r(z_0)=\frac{\sigma_0(z_0)\tau_\infty(z_0)-\tau_0(z_0)\sigma_\infty(z_0)}{\tau_0'(z_0)\tau_\infty(z_0)-\tau_0(z_0)\tau_\infty'(z_0)}=\delta(z_0)/W(z_0)$$

where $W(z)$ is the Wronskian

$$W(z)=\begin{vmatrix}\tau_0'(z) & \tau_\infty'(z)\\ \tau_0(z) & \tau_\infty(z)\end{vmatrix}.$$

We should note that the functions $\delta(z)$ and $W(z)$ depend on the choice of the basis $[\sigma_0;\tau_0]$ and $[\sigma_\infty;\tau_\infty]$ in the space $\mathscr{S}$. If we pass from one basis to another by some linear transformation, these determinants are both multiplied by the non-zero determinant of that transformation. Their ratio, however, is independent of the choice of basis, as it should be, since $r(z)$, the eventual residue, has a meaning independent of the basis.

Our analysis so far has shown that there exist two widely different possibilities for the interpolation problem. Either the corresponding linear problem admits a solution of lower degree, in which case the interpolation problem has either a unique solution or none at all, or there are no solutions of lower degree and the corresponding interpolation problem has a whole family $f_t(z)$ of solutions. It is clear that a criterion is needed to distinguish between these two cases. To obtain such a criterion we will rewrite the equations of the linear problem in terms of divided differences.

The data of the Cauchy Interpolation Problem is the set of N numbers f_{ij} specifying the values of the derivatives of the solution function

$f(z)$. These derivatives being known, all related divided differences of the function are also known. Thus, if

$$S=\{z_1, z_1, \ldots, z_1, z_2, \ldots, z_l\}$$

is the set of N numbers formed by taking each z_k ν_k+1 times, then all divided differences of any solution to the interpolation problem corresponding to any subset of S are determined. We may therefore think of the data of the problem as the specification of all these divided differences. If m points of S are given in the form $z_1, z_2, \ldots, z_m$ we write the corresponding number $[z_1, z_2, \ldots, z_m]$ without a subscript, since we do not know that the Cauchy Interpolation Problem necessarily has a solution. However, there always will exist a polynomial $F(z)$ of degree at most N satisfying the N equations

$$F^{(j)}(z_i)=f_{ij}$$

and for which the divided differences are those of the interpolation problem, i.e.,

$$[z_1, z_2, \ldots, z_m]_F=[z_1, z_2, \ldots, z_m].$$

Let $\lambda_1, \lambda_2, \ldots, \lambda_{2n}$ be an ordering of S. We expand the polynomials $\sigma(z)$ and $\tau(z)$ in terms of the Newtonian Interpolation Polynomials $P_i(z)$ associated with this sequence. Accordingly

$$\sigma(z)=\sum a_i P_i(z) \quad \text{where } a_i=[\lambda_1, \lambda_2, \ldots, \lambda_{i+1}]_\sigma$$

and

$$\tau(z)=\sum b_j P_j(z) \quad \text{where } b_j=[\lambda_1, \lambda_2, \ldots, \lambda_{j+1}]_\tau .$$

We write $\sigma(z)=\tau(z)f(z)$ and take the divided differences of each side of this equation, and from the Leibnitz rule we obtain the $2n$ linear equations

$$[\lambda_1 \ldots \lambda_m]_\sigma=\sum_{l=0}^{m-1}[\lambda_1 \ldots \lambda_{l+1}]_\tau[\lambda_{l+1} \ldots \lambda_m]_f .$$

These equations may also be written

$$a_j=\sum b_k[\lambda_{k+1}, \lambda_{k+2}, \ldots, \lambda_{j+1}] \quad j=0,1, \ldots, 2n-1,$$

where we now omit the subscript. These $2n$ equations give rise to equally many linear conditions on the space $\mathscr{M}$ of pairs of polynomials $[\sigma; \tau]$ of degree at most n. Since that space has dimension $2n+2$ it follows that there exists a subspace $\mathscr{S}'$ of solutions to these equations of dimension at least 2. We must show that $\mathscr{S}'$ and $\mathscr{S}$ are the same, i.e., that the new equations and the old equations are equivalent systems, and therefore that the new equations equally well represent the linear problem.

For this purpose we let $F(z)$ be the polynomial of degree at most N introduced above, and which is a solution to a related interpolation problem. The new set of equations is precisely the assertion that the functions $\sigma(z)$ and $\tau(z)F(z)$ have the same divided differences relative to the $2n$ sets $\lambda_1, \lambda_2, \lambda_3, \ldots, \lambda_m$. Since these divided differences are enough to determine the divided difference relative to any subset of S, the two functions have the same divided difference relative to any such subset. On the other hand, the initial equations for the Linear Problem were exactly the assertion that the functions $\sigma(z)$ and $\tau(z)F(z)$ had the same values at every interpolation point z_i and so did their derivatives of order up to v_i. This also means that the divided differences relative to any subset of S are the same. Thus a pair $[\sigma;\tau]$ is a solution to one set of equations only if it is also a solution to the other.

Suppose now there exists a solution $[\sigma;\tau]$ of lower degree. This means that the equations have a solution with $a_j = b_j = 0$ for $j \geqq n$. The n equations obtained for $j \geqq n$ are a homogeneous system in the n coefficients b_k:

$$0 = \sum b_k [\lambda_{k+1}, \lambda_{k+2}, \ldots, \lambda_{j+1}] \qquad j = n, n+1, \ldots, 2n-1$$

and it is evident that the determinant of this system must vanish. Conversely, if that determinant vanishes, there exists a non-trivial solution to the last n equations with $b_n = 0$. The first n equations then determine a system of coefficients a_k with $a_n = 0$, and it follows that there exists a polynomial pair $[\sigma;\tau]$ of lower degree in the solution space $\mathscr{S}$.

We can therefore distinguish the case in which the linear problem has solutions of lower degree by considering the determinant of the matrix which follows.

$$\begin{bmatrix} [\lambda_1 \lambda_2 \ldots \lambda_{n+1}] & [\lambda_2 \lambda_3 \ldots \lambda_{n+1}] & \cdots & [\lambda_n \lambda_{n+1}] \\ [\lambda_1 \lambda_2 \ldots \lambda_{n+2}] & [\lambda_2 \lambda_3 \ldots \lambda_{n+2}] & \cdots & [\lambda_n \lambda_{n+1} \lambda_{n+2}] \\ \vdots & \vdots & & \vdots \\ [\lambda_n \quad \ldots \lambda_{2n}] & [\lambda_2 \quad \ldots \lambda_{2n}] & \cdots & [\lambda_n \quad \ldots \lambda_{2n}] \end{bmatrix}.$$

We set

$$\xi_1 = \lambda_{n+1}, \xi_2 = \lambda_{n+2}, \xi_3 = \lambda_{n+3}, \ldots, \xi_n = \lambda_{2n}$$

and

$$\eta_1 = \lambda_n, \eta_2 = \lambda_{n-1}, \eta_3 = \lambda_{n-2}, \ldots, \eta_n = \lambda_1$$

and invert the order of the columns of the matrix to obtain the extended Loewner matrix

$$L_e = \begin{bmatrix} [\xi_1, \eta_1] & [\xi_1, \eta_1, \eta_2] & [\xi_1, \eta_1, \eta_2, \eta_3] & \cdots & [\xi_1, \eta_1, \ldots, \eta_n] \\ [\xi_1, \xi_2, \eta_1] & [\xi_1, \xi_2, \eta_1, \eta_2] & & & \\ [\xi_1, \xi_2, \xi_3, \eta_1] & & & & \\ \vdots & & & & \\ [\xi_1, \xi_2, \ldots, \xi_n, \eta_1] & & & & [\xi_1, \xi_2, \ldots, \xi_n, \eta_1, \ldots, \eta_n] \end{bmatrix}.$$

We can sum up our study of the interpolation problem for $N=2n$ as follows. If $\det L_e$ is not zero, there exists a one-parameter family

$$f_t(z)=\frac{\sigma_0(z)+t\,\tau_\infty(z)}{\tau_0(z)+t\,\tau_\infty(z)}$$

of solutions to the interpolation Problem, where certain exceptional values of t, at most l in number, must be excluded. All these solutions are of degree n. If $\det L_e=0$, then either there exists no solution, or there exists exactly one and the degree of this solution is smaller than n.

These circumstances lead to the following useful result.

Theorem I. *Let $f(z)$ be a rational function of degree d. Choose $2n$ numbers none of which is a pole of $f(z)$ and write n of them as $\xi_1,\xi_2,\ldots,\xi_n$ and the remaining n as $\eta_1,\eta_2,\ldots,\eta_n$. Form the extended Loewner matrix associated with this choice of the points ξ and η and the function $f(z)$. Then if $n=d$ the determinant is not zero, while it vanishes if $n>d$.*

Proof. The function $f(z)$ is the ratio of two relatively prime polynomials $\sigma(z)$ and $\tau(z)$, each of degree at most d and at least one of them is of degree exactly d. Moreover, $f(z)$ is a solution to the Cauchy Interpolation Problem associated with this Loewner matrix. It follows that if $n=d$ that problem has a solution of degree d. It does not have a solution of lower degree since then $\sigma(z)$ and $\tau(z)$ would have a common zero. Hence the determinant is not zero. If $n>d$ the corresponding interpolation problem has a solution of lower degree, viz. $f(z)$, and therefore the determinant vanishes.

It should be emphasized that the extended Loewner matrix contains all of the data of the Interpolation Problem, and the vanishing or non-vanishing of its determinant is a property of the Interpolation Problem and has nothing to do with the way in which the set S is divided into n numbers ξ_i and n numbers η_j. Different divisions of the set and different orderings may give rise to different values of the determinant, but all will be non-zero in such a case. Accordingly, the numbers ξ_i and η_j may always be taken in the most convenient fashion.

At this point we introduce the concept of an associated determinant. For every subset of S of even cardinal we choose a representation in the form ξ,η and compute the corresponding extended Loewner matrix and its determinant. Such a determinant will be said to be associated with the subset. It will be of order k if the cardinal of the subset is $2k$. From the previous theorem it is now clear that a necessary condition for the existence of a solution of degree d to the Cauchy Interpolation Problem is that all associated determinants of order d are non-zero, while all associated determinants of higher order vanish. We shall presently see that this condition is almost, but not quite, sufficient.

Lemma 4. *Let $N=2n$ and suppose that the associated determinant of order n vanishes, but that all associated determinants of order $n-1$ are non-zero. Then there exists a unique solution to the Cauchy Interpolation Problem and it is of degree exactly $n-1$.*

Proof. Since $\det L_e=0$ we know that if a solution exists it is of lower degree and is unique. In view of the theorem above we see that its degree must be exactly $n-1$. Hence what must be proved is that there actually exists a solution. For this purpose we consider the equations determining the polynomial $\tau(z)$. We have

$$\sum_{j=1}^{n} a_{ij} b_{n-j}=0$$

where we have put $a_{ij}=[\xi_1,\xi_2,\ldots,\xi_i,\eta_1,\eta_2,\ldots,\eta_j]$. The determinant of this system of equations is zero and the solution is determined up to a multiplicative constant. Let A_{ij} be the minor corresponding to a_{ij}; now $b_{n-j}=(-1)^{j+n}A_{nj}$ is a solution to the system. In particular then $b_0=A_{nn}$ cannot be 0 since A_{nn} is the determinant associated with the set obtained from S by striking out the numbers ξ_n and η_n. Accordingly

$$\tau(z)=\sum b_k P_k(z)$$

and $b_0=\tau(\lambda_1)=\tau(\eta_n)$ is not 0. Since we can arrange the description of S in terms of the ξ_i and η_j in any way we like, we can make η_n equal to an arbitrary interpolation point z_i. It follows that the corresponding polynomial $h_0(z)$ is a non-zero constant. Hence there is a solution to the interpolation problem.

Theorem II. *Let d be an integer satisfying the inequality $2d\geqq\max_i v_i$ and suppose every associated determinant of order d is non-zero, while all associated determinants of higher order vanish. Then there exists a solution to the interpolation problem, evidently of degree d.*

Proof. If $d=n$ the problem is known to have an infinite family of solutions. For $d<n$ we consider any subset S' of S with cardinal $2d+2$. The previous lemma guarantees that the interpolation problem associated with S' has a solution. Suppose two sets of cardinal $2d+2$ have $2d+1$ points in common. Since $2d+1$ conditions are enough to determine a rational function of degree d, both sets are associated with the same rational solution. We can pass from any subset of S of cardinal $2d+2$ to any other such set by successively adding and taking away one point at a time. Thus all subsets of the cardinal $2d+2$ are associated with the same solution $f(z)$ of degree d. For any interpolation point z_i the set consisting of z_i taken v_i+1 times has a cardinal no

larger than $2d+1$ and so $f(z)$ satisfies all of the equations of the interpolation problem.

The condition $2d \geqq \max \nu_i$ in the previous theorem is a necessary one as the following example shows. We consider the interpolation problem with just one interpolation point z_1 and suppose that the function $f(z)$ and its derivatives up to and including the fifth derivative are prescribed. We will have $N=6$ and $\nu_1=5$ and we require.

$$f_{10}=f_{11}=f_{12}=f_{13}=0$$

and put $a=f_{14}/4!$ and $b=f_{15}/5!$. The extended Loewner matrix of order 3 has the form

$$\begin{bmatrix} 0 & 0 & 0 \\ 0 & 0 & a \\ 0 & a & b \end{bmatrix}$$

and all the associated determinants vanish. Hence $d=0$. It is easy to see that the only possible solution to the interpolation problem is $f(z)=0$ identically, and this will be the case if and only if $a=b=0$.

Our extensive study of the interpolation problem for even values of N now makes possible the same results for odd N.

Theorem III. *The statement of Theorem II holds also when N, the number of conditions of the interpolation problem, is odd.*

Proof. We suppose $N=2n+1$ and first dispose of the special case when there is only one interpolation point. We will have $2d \geqq \nu_1 = N-1$ and so the problem of order $2n$ obtained by omitting the condition of highest order will have a non-zero determinant. There will then be a family $f_t(z)$ of solutions to that problem and precisely one of them will satisfy the final equation involving the derivative of order ν_1.

Since we may now suppose that the points z_1 and z_l are distinct, we consider the two interpolation problems of order $2n$ obtained by omitting the highest order condition at each of these points. If $d<n$, each of these problems admits a unique solution, of order at most $n-1$, and since $2n-1$ conditions are enough to determine a solution, these solutions are the same, since the problems have $2n-1$ conditions in common. Thus the interpolation problem has a solution in this case as well.

We must finally consider the case when $d=n$. Here we let $f_t(z)$ be the family of solutions of the problem obtained by omitting the highest order condition at z_1. We then select t so that $f_t(z)$ satisfies the omitted condition. We will show that this function is not an exceptional solution, and hence that it is the solution to our problem. If the solution were exceptional relative to some point, say z_j, then we write

$$\sigma_t(z) = (z - z_j)\hat{\sigma}(z)$$

and

$$\tau_t(z) = (z - z_j)\hat{\tau}(z)$$

to obtain a polynomial pair $[\hat{\sigma}; \hat{\tau}]$ of lower degree satisfying every equation of the linear problem corresponding to the interpolation problem of order $2n$ obtained by omitting the highest order condition at z_j. Thus the corresponding n-th order determinant must vanish, contrary to hypothesis. This completes the proof of the theorem.

We conclude this chapter with two useful identities associated with the interpolation problem when $N = 2n$ is even. For this purpose we suppose that A is the Loewner matrix associated with the problem, that is, that

$$a_{ij} = [\xi_1, \xi_2, \ldots, \xi_i, \eta_1, \ldots, \eta_j]$$

and we write the minor of a_{ij} as A_{ij}. Let the polynomials $p_i(z)$ be the Newtonian interpolation polynomials associated with the sequence $\xi_1, \xi_2, \ldots$ while the $q_j(z)$ are similarly associated with the sequence $\eta_1, \eta_2, \ldots$. We shall also suppose that $\det A$ is not zero. Let z_0 be any complex number such that $T(z_0)$ is not exceptional; this hypothesis only excludes a finite set of possible values for z_0. If $t = T(z_0)$ the function $f_t(z)$ corresponding has a pole at z_0 and we may write it displaying that pole as follows:

$$f_t(z) = h(z) + \frac{r(z_0)}{z - z_0}$$

where $h(z)$ is a rational function of degree smaller than n. It follows that the extended Loewner matrix computed for the function $h(z)$ has a determinant which vanishes. The corresponding matrix elements are

$$[\xi_1, \xi_2, \ldots, \xi_i, \eta_1, \ldots, \eta_j]_h = a_{ij} + r(z_0)/p_i(z_0)\,q_j(z_0).$$

We expand the determinant in the equation

$$0 = \det[a_{ij} + r/p_i q_j]$$

to obtain

$$0 = \det A + r(z_0) \sum\sum \frac{(-1)^{i+j} A_{ij}}{p_i(z_0)\,q_j(z_0)}.$$

Now put

$$\chi(z) = \sum\sum \frac{(-1)^{i+j} A_{ij}}{p_i(z)\,q_j(z)}$$

to find

$$r(z_0) = \frac{-\det A}{\chi(z_0)} = \delta(z_0)/W(z_0).$$

Finally, since $\delta(z)\chi(z)$ is a polynomial, we infer the identity

$$W(z)=\frac{\delta(z)\chi(z)}{-\det A}.$$

We may suppose that a basis has been selected in the solution space $\mathscr{S}$ so that $\delta(z)=\Delta(z)$, and this polynomial is of course the same as $p_n(z)q_n(z)$. Now

$$(-\det A)W(z)=\Delta(z)\chi(z)=\sum\sum(-1)^{i+j}A_{ij}\frac{p_n(z)}{p_i(z)}\frac{q_n(z)}{q_j(z)}$$

and the ratio $q_n(z)/q_j(z)$ is of degree $n-j$ and has zeros at the points $\eta_{j+1},\eta_{j+2},\ldots,\eta_n$; it is therefore the polynomial which we have earlier called $P_{n-j}(z)$, when we set up the equations determining $\tau(z)$. In a similar way we define $Q_{n-i}(z)=p_n(z)/p_i(z)$ to obtain

$$\det A\, W(z)=-\sum\sum(-1)^{i+j}A_{ij}P_{n-j}(z)Q_{n-i}(z)$$

and this polynomial is the determinant of the following matrix.

$$\begin{bmatrix} 0 & P_{n-1}(z) & P_{n-2}(z) & \ldots & P_0(z)\\ Q_{n-1}(z) & a_{11} & a_{12} & \ldots & a_{1n}\\ Q_{n-2}(z) & a_{21} & a_{22} & \ldots & a_{2n}\\ \vdots & \vdots & \vdots & & \vdots\\ Q_0(z) & a_{n1} & a_{n2} & \ldots & a_{nn}\end{bmatrix}$$

Let us turn now to a special case: we shall suppose that $\det A=0$ but that all associated determinants of order $n-1$ are non-zero. In this case, according to Lemma 4, the interpolation problem admits a unique solution, of degree exactly $n-1$. The polynomial pair $[\sigma;\tau]$ is determined only up to a multiplicative constant, and we will determine $\tau(z)$ as in the proof of that lemma, viz. by the condition that b_0 is the minor A_{nn} of the Loewner matrix. Now consider the auxiliary function of two variables:

$$\chi(x,y)=\sum\sum\frac{(-1)^{i+j}A_{ij}}{p_i(x)q_j(y)}.$$

We have

$$p_n(x)q_n(y)\chi(x,y)=\sum\sum(-1)^{i+j}A_{ij}P_{n-j}(y)Q_{n-i}(x)$$

and this quantity may be written

$$\sum Q_{n-i}(x)\sum(-1)^{i+j}A_{ij}P_{n-j}(y).$$

Now when $i=n$ the second sum is exactly $\tau(y)$ according to the proof of Lemma 4, and when i is different from n the coefficients A_{ij} are proportional to A_{nj}. Our polynomial therefore takes the form

$$\sum C_i Q_{n-i}(x)\tau(y)$$

and so $\tau(y)$ evidently divides the polynomial $p_n(x)q_n(y)\chi(x,y)$. Consider next the very same interpolation problem, where we choose the ξ_i and η_j differently, interchanging ξ_i and η_i for every i. This corresponds to the passage from A to its transpose and to the substitution of Q's for P's. The polynomial $\tau(z)$ is still associated with this reformulated problem. We interchange x and y to obtain

$$\begin{aligned} p_n(y)q_n(x)\chi(y,x) &= \sum\sum(-1)^{i+j}A_{ij}P_{n-j}(x)Q_{n-i}(y) \\ &= \sum\sum(-1)^{i+j}A^*_{ji}Q^*_{n-j}(x)P^*_{n-i}(y) \\ &= p^*_n(x)q^*_n(y)\chi^*(x,y) = p_n(y)q_n(x)\chi^*(x,y). \end{aligned}$$

Here the stars indicate the reformulated problem.

We infer that $\chi^*(x,y)=\chi(y,x)$ where $\chi^*(x,y)$ is the function associated with the other choice of the points ξ_i and η_j. Since $\tau(y)$ divides $p^*_n(x)q^*_n(y)\chi^*(x,y)$ it also divides $p_n(y)q_n(x)\chi(y,x)$ and it finally follows that $\tau(x)$ divides $p_n(x)q_n(y)\chi(x,y)$. Thus $\tau(x)\tau(y)$ divides that polynomial. We next set $x=y$ to obtain

$$\Delta(x)\chi(x) = C\tau(x)^2$$

for some non-zero constant C, since the degrees of the polynomials on either side are the same.

Chapter XIII. Interpolation by Pick Functions

In this chapter we study interpolation problems which admit solutions which are rational Pick functions, real on the real axis. We first give an instructive alternate proof of Pick's theorem.

Pick's Theorem. *Let Z be a finite set of n points in the upper half-plane*

$$z_1, z_2, \ldots, z_n$$

and

$$w_1, w_2, \ldots, w_n$$

be equally many (not necessarily distinct) complex numbers. If

$$K_{ij} = K(z_i, z_j) = \frac{w_i - \overline{w}_j}{z_i - \overline{z}_j}$$

is a positive matrix, then there exists a rational function $\varphi(z)$ in the Pick class of degree at most n, real on the real axis and satisfying the n equations $\varphi(z_i) = w_i$.

Proof. We first prove the theorem under the additional hypothesis that $\det K > 0$. From the Schwartz reflection principle it follows that a function $\varphi(z)$ satisfying the conditions of the theorem must also satisfy the n equations

$$\varphi(\overline{z}_i) = \overline{w}_i$$

and we are therefore led to consider the interpolation problem associated with the set Z and its reflection in the lower half-plane. We will set

$$z_i = \xi_i \quad \text{and} \quad \overline{z}_j = \eta_j$$

and pass to the extended Loewner matrix L_e associated with this interpolation problem. In Chapter III we observed how the determinants of K and L_e were related:

$$\det K = \det L_e \prod_{i>j} |z_i - z_j|^2$$

and we now infer that $\det L_e$ is positive. We can argue similarly with the interpolation problem of lower order obtained from this one by considering only the points $z_1, z_2, \ldots, z_k$ and the corresponding values of w_i to infer that the matrix obtained from L_e by omitting the final $n-k$ rows and columns has a positive determinant. The criterion for positivity proved in Chapter I then guarantees that L_e is a positive matrix. Moreover, the interpolation problem associated with L_e has an infinite family $f_t(z)$ of solutions, and our problem is to make sure that at least one of these is a Pick function, real on the real axis.

We next note that if $[\sigma; \tau]$ is a solution to the corresponding linear problem then for every i

$$\sigma(z_i) = w_i \tau(z_i)$$

and also

$$\sigma(\bar{z}_i) = \bar{w}_i \tau(\bar{z}_i).$$

Taking complex conjugates in the last equation leads to

$$\overline{\sigma(\bar{z}_i)} = w_i \overline{\tau(\bar{z}_i)}$$

and we infer that the polynomials $\overline{\sigma(\bar{z})}$ and $\overline{\tau(\bar{z})}$ also satisfy the equations of the linear problem. This makes it clear that we can select in the space $\mathscr{S}$ of solutions to the linear problem a basis $[\sigma_0; \tau_0]$ and $[\sigma_\infty; \tau_\infty]$ consisting of polynomials with real coefficients. We then pass to the system

$$\begin{aligned} \sigma_t(z) &= \sigma_0(z) + t\,\sigma_\infty(z), \\ \tau_t(z) &= \tau_0(z) + t\,\tau_\infty(z) \end{aligned}$$

where we take only real values of the parameter t. These polynomials will therefore have real coefficients and will be real on the real axis, and so will be the corresponding functions $f_t(z)$. We also note that the Wronskian will be a polynomial with real coefficients.

Suppose that the basis is so normalized that $\delta(z) = \Delta(z)$, let $W(z)$ be the Wronskian and A the extended Loewner matrix. Following the calculations at the end of the previous chapter we may write $\det A\, W(z)$ as the determinant of the following matrix:

$$\begin{bmatrix} 0 & P_{n-1}(z) & P_{n-2}(z) & \ldots & P_0(z) \\ Q_{n-1}(z) & a_{11} & a_{12} & \ldots & a_{1n} \\ Q_{n-2}(z) & a_{21} & a_{22} & \ldots & a_{2n} \\ \vdots & \vdots & \vdots & & \vdots \\ Q_0(z) & a_{n1} & a_{n2} & \ldots & a_{nn} \end{bmatrix}$$

Note that when only real values of z are taken the matrix above is symmetric since $P_i(x) = \overline{Q_i(x)}$ for real x. From the criterion for positivity given in Chapter I it follows that this matrix is a positive matrix if its determinant is $\geqq 0$. However, the matrix cannot be a positive matrix since it has a 0 in the upper left hand corner, and, if it were positive, the whole top row would have to vanish, contradicting $P_0(x)=1$. We infer that the determinant in question is strictly negative, whence $W(x)<0$ for real x. It follows that the polynomials $\tau_t(z)$ have only simple zeros on the real axis. Since for all real x $\Delta(x)=\Pi|x-z_i|^2>0$, the eventual residue $r(x)$ is negative on the real axis. Accordingly, if we can show that for some value of t, say $t=0$, the polynomial $\tau_0(z)$ has no non-real zeros, then the corresponding $f_0(z)$ will be real on the real axis and will have all of its singularities there, and these will be simple poles with negative residues. Thus $f_0(z)$ will be a Pick function. Indeed, we can go farther and assert that every $f_t(z)$ is a Pick function, because it is easy to see that every $\tau_t(z)$ has only real zeros. The zeros of these polynomials are surely continuous functions of t and these functions are always real, since a non-zero root can appear only in a conjugate pair which comes into existence after two real roots coincide on the axis. But such multiple zeros can never occur since $W(x)$ never vanishes. We also infer that for no real t is $f_t(z)$ an exceptional solution.

Choose a function $\psi(z)$ in the Pick class of the form

$$\psi(z)=1+\sum_{i=1}^{n}\frac{m_i}{\lambda_i-z}$$

where $\lambda_1=0$. Consider the interpolation problem associated with the set Z and the corresponding values

$$s\,w_i+(1-s)\,\psi(z_i)$$

where s is fixed in the closed interval $[0,1]$. Clearly $s=1$ refers to the problem we are studying, while $s=0$ refers to a problem having the function $\psi(z)$ among its solutions. The Pick matrix associated with s is the matrix

$$s\,K_{ij}+(1-s)\,H_{ij}$$

where, of course, H_{ij} is the Pick matrix associated with $\psi(z)$ and Z. As a convex combination of positive matrices we see that this matrix is a positive matrix, indeed, positive definite. Thus for each s in the interval the corresponding Wronskian $W_s(x)$ has no zero on the real axis. This function clearly depends continuously on s. Now let $\tau(z,s)$ be that polynomial associated with the s-problem which is so normalized that $\tau(0,s)=0$ and $\tau'(0,s)=1$. This function depends continuously on s since it is uniquely determined by these two equations and the $n-1$ homogeneous equations of the interpolation problem, the coefficients in those equations depending continously on s. For no value of s can $\tau(z,s)$

have a multiple real zero, while for $s=0$ the corresponding polynomial has n real and simple zeros, namely the poles of $\psi(z)$. It follows that $\tau(z,1)$ has no non-real zeros, completing the proof of the theorem when $\det K$ is not 0.

When that determinant vanishes we pass instead to a related problem, taking a small positive ε and the new values $w_i+\varepsilon\log z_i$. Let $\varphi_\varepsilon(z)$ be a solution to this related problem which surely is associated with a positive definite Pick matrix. As ε converges to 0 a suitable sequence from this family of functions converges to a solution of the original interpolation problem, thereby completing the proof.

The principal theorem of this chapter is concerned with the Cauchy interpolation problem when all of the data of that problem are real and where the solution is to be a Pick function. Thus the interpolation points are on the real axis

$$x_1<x_2<\cdots<x_l$$

and the corresponding numbers f_{ij} are also real. In this case the whole analysis of the previous chapter must be taken relative to the field of real numbers rather than the field of complex numbers. The polynomials $\sigma(x)$ and $\tau(x)$ will have real coefficients and the functions $f_t(x)$, $W(x)$, etc., will all be real on the real axis. The results of the previous chapter will all be valid in this more restricted interpretation.

Because of our interest in Pick functions another convention is required. Given a subset of S consisting of $2m$ numbers we shall define the determinant associated with this subset more explicitly than was necessary in Chapter XII. We shall write the set in the form

$$\xi_1\leqq\eta_1\leqq\xi_2\leqq\cdots\leqq\xi_m\leqq\eta_m$$

and take the determinant of the extended Loewner matrix relative to this representation of the set in terms of ξ_i and η_j. Such determinants are surely nonnegative if there exists a solution to the interpolation problem which is a Pick function, real on the real axis and having no poles in some interval containing the interpolation points.

Note that the following theorem is valid even when the number of conditions of the Cauchy problem is odd.

Theorem I. *Suppose* $2d\geqq\max_i v_i$ *and that every associated determinant of order* $\leqq d$ *is strictly positive, while all such determinants of higher order vanish. Then there exists a solution which is a Pick function having its poles outside the closed interval* $[x_1,x_l]$. *The solution is evidently of degree d.*

Proof. The hypothesis of the theorem is strong, and the proof is elementary but long. We argue by induction on $N=\sum(v_i+1)$, the number of conditions of the problem. We first suppose that N is even, putting $N=2n$ and assume the theorem true for $N=2n-1$. Later we will pass from $N=2n$ to $2n+1$, completing the proof. We may suppose that $n>1$, since the case $n=1$ is trivial.

If $d<n$ the results of the previous chapter show the existence of a unique rational solution $f(x)$ of degree d. On the other hand, if we omit one of the conditions of the problem, passing therefore to a new problem with $2n-1$ conditions, the inductive hypothesis guarantees that the new problem has a unique solution $g(x)$ which is a Pick function of degree d. Evidently $f(x)$ and $g(x)$ coincide, and it is now easy to check that this Pick function has its poles outside the interval $[x_1, x_l]$. We may therefore take $d=n$ and we know that there exists a system $f_t(x)$ of solutions to the interpolation problem. We claim that the exceptional solutions in this problem are distinct: if $f_0(x)$ is exceptional relative to the point x_i, then it is not exceptional relative to any other point. This is a consequence of the fact that the polynomials $\sigma_0(x)$ and $\tau_0(x)$ are uniquely determined up to a common constant multiplier by the $2n$ equations of the linear problem and the further condition $\tau_0(x_i)=0$. Now, from the inductive hypothesis, if we consider the interpolation problem with $2n-1$ conditions obtained by omitting the condition of highest order at x_i, that problem has a unique solution which is representable as a ratio $\hat{\sigma}(x)/\hat{\tau}(x)$ where the polynomials in question are relatively prime and of degree at most $n-1$. It follows that the pair $(x-x_i)\hat{\sigma}(x)$ and $(x-x_i)\hat{\tau}(x)$ have only x_i as a common zero and satisfy all of the $2n$ equations of the linear problem. We infer that $\tau_0(x)=(x-x_i)\hat{\tau}(x)$ and therefore that $\tau_0(x)$ does not vanish at any other interpolation point. The fact that the exceptional solutions are distinct will be important often in the sequel.

Our proof depends in an essential way on the fact that a basis may be selected in $\mathscr{S}$ in such a way that $\delta(x)=\Delta(x)$ and the corresponding Wronskian $W(x)$ is negative on the real axis. For if this is shown, the argument for $N=2n$ is easily completed as follows. Let $f_0(x)$ be the exceptional solution associated with the first interpolation point x_1 which is to the left of the others. By virtue of the inductive hypothesis, $f_0(x)$ is a solution to the interpolation problem of order $2n-1$ obtained by omitting the condition of highest order at x_1. It follows that $f_0(x)$ is a Pick function with its poles outside the interval $[\eta_1, \xi_n]$. If η_1 is not equal to ξ_1 we make use of the fact that $W(x)$ and $\Delta(x)$ are both negative in the interval (ξ_1, η_1) to infer that the eventual residue is positive there. We see that $f_0(x)$ has no pole in that interval, since the residues of a Pick function are always negative. Hence $f_0(x)$ has no pole in the closed interval $[x_1, x_l]$. For some values of t very close to 0, then, $f_t(x)$ will have all of its poles outside that interval, and these poles will all have negative residues. Such functions are then Pick functions satisfying the assertions of the theorem.

Our argument now reduces to showing $W(x)$ negative. In the special case that the problem is so symmetric that $\xi_i = \eta_i$ for every i we can argue as in the proof of Pick's theorem, writing $W(x)$ as the determinant of a symmetric matrix which cannot be a positive matrix, although the extended Loewner matrix is a positive matrix. We will infer that $W(x) < 0$ on the axis. In general, no such symmetry hypothesis is available and the points ξ_i and η_i are distinct. Our argument will therefore be more complicated. It should be noted, however, that when there is only one interpolation point then the ξ and η obviously do coincide. We should also note that the representation of $W(x)$ as a determinant displays the continuous dependence of $W(x)$ on the data of the problem, an obvious but important fact.

In order to make $W(x)$ negative it is enough to exhibit a basis $[\sigma_0; \tau_0]$ and $[\sigma_\infty; \tau_\infty]$ in $\mathcal{S}$ such that $\delta(x) = \Delta(x)$ and for which the polynomial $\tau_\infty(x)$ has n real and simple zeros λ_i at which the Wronskian is negative. For in this case we form

$$T(x) = -\tau_0(x)/\tau_\infty(x)$$

and compute the residue at the pole λ_i. This is

$$\lim_{x \to \lambda_i} \frac{-(x-\lambda_i)\tau_0(x)}{(x-\lambda_i)\tau'_\infty(\lambda_i) + (x-\lambda_i)^2 H(x)} = \frac{-\tau_0(\lambda_i)}{\tau'_\infty(\lambda_i)} = W(\lambda_i)/\tau'_\infty(\lambda_i)^2 < 0.$$

Accordingly $T(x)$ has simple real poles with negative residues and is therefore a Pick function. We compute its derivative to find

$$T'(x) = \frac{-\tau_\infty(x)\tau'_0(x) + \tau'_\infty(x)\tau_0(x)}{\tau_\infty(x)^2} = -W(x)/\tau_\infty(x)^2$$

and now infer that $W(x)$ is negative, since $T(x) = \beta + \sum (m_i/(\lambda_i - x))$ cannot have a zero derivative on the real axis:

$$T'(x) = \sum \frac{m_i}{(\lambda_i - x)^2} > 0.$$

Now choose t so that $f_t(x)$ is an exceptional solution associated with the point η_1. A moments reflection makes it clear that this function has its poles outside the interval $[x_1, x_l]$ and that the residues at those poles are negative, whence $W(x)$ is negative at those poles because $\Delta(x)$ is positive there. We shall assume that there are in fact $n-1$ such poles, i.e., that $f_t(x)$ does not have a pole at infinity. If this were the case we would alter the data of the problem slightly, obtaining a problem where this assumption is valid. We will show that $W(x)$ is not positive in a small neighborhood of the interpolation point η_1: it will then follow that for s very close to t the polynomial $\tau_s(x)$ is such that $W(x)$ is negative at its n simple real zeros, and this, as we have seen, is enough to guarantee $W(x)$ negative on the real axis.

Accordingly, for s very close to $T(\eta_1)=t$ we have

$$f_s(z)=h(z)+\frac{r(x)}{z-x}$$

where x is close to η_1 and is a zero of $\tau_s(z)$. We will show that $r(x)$ and $\Delta(x)$ have opposite signs, thereby guaranteeing that $W(x)$ is negative. The function $h(z)$ above is a Pick function of degree exactly $n-1$ having its poles outside the interpolation interval $[x_1, x_l]$. Let B be the extended Loewner matrix associated with $h(z)$ and the points ξ, η of our problem. We know $\det B=0$ since $h(z)$ is of lower degree. Similarly, let A be the extended Loewner matrix of our interpolation problem. Now

$$0<\det A=\det[b_{ij}+r/p_i q_j]=\det B-r(x)\sum\sum\frac{(-1)^{i+j}B_{ij}}{p_i(x)\,q_j(x)}$$

where, as before, B_{ij} is the determinant of the matrix obtained by omitting the i-th row and j-th column from B. In the notation of the previous chapter this may be written

$$0<-r(x)\,\chi(x)=(-r(x)/\Delta(x))\,\Delta(x)\,\chi(x)$$

and because B is of rank exactly $n-1$ we have from the remarks at the end of that chapter

$$0<(-r(x)/\Delta(x))\,C\,\hat{\tau}(x)^2$$

where $h(z)=\hat{\sigma}(z)/\hat{\tau}(z)$, the polynomials $\hat{\sigma}$ and $\hat{\tau}$ being relatively prime. We shall presently show that C is positive and infer that $\Delta(x)$ and $r(x)$ have opposite signs, completing this stage of the proof.

To make sure that C is positive, consider an x outside the interval $[x_1, x_l]$ which is not a pole of $h(z)$ and take some negative r to form the Pick function

$$h(z)+\frac{r}{z-x}$$

which is of course associated with an extended Loewner matrix having a positive determinant. Accordingly

$$0<\det[b_{ij}+r/p_i q_j]=-r\,\chi(x)=(-r/\Delta(x))\,C\,\hat{\tau}(x)^2$$

and since r is negative and $\Delta(x)$ positive, we find $C>0$. This completes the proof in the case $N=2n$, however, before passing to the induction from $2n$ to $2n+1$ we should perhaps discuss one point in the previous argument which may be obscure. We treated a particular exceptional solution $f_t(x)$ and wanted to assume that this function did not have a pole at infinity. This meant that the corresponding $\tau_t(x)$ was in fact of degree n and not of lower degree. Now the coefficient of z^n in that polynomial may be taken to be a determinant which depends continuously on the data of the interpolation problem, since that polynomial is a solution of $n-1$ homogeneous equations given with the problem itself, as well as a further homogeneous equation of the form $\tau(x_i)=0$. The determinant in question is a polynomial in the data f_{ij} of the problem and does not vanish identically. From this it follows that a very slight change in the data of the problem gives rise to one where the corresponding exceptional solution has no pole at infinity. But since $W(x)$ depends continuously on the data, we can infer that $W(x)\leqq 0$ in any case.

We suppose next that $N=2n+1$ and again argue by induction. Let us note that Theorem III of the previous chapter guarantees the existence of a unique solution to the interpolation problem, and our task is merely to show that the solution in question is a Pick function, having its poles outside the interpolation interval. Let us first dispose of the special case where there is only one interpolation point. We will have $2d\geqq v_1=N-1=2n$, and so the problem obtained by omitting the highest order condition will have a non-zero determinant. The family of solutions $f_t(x)$ corresponding will consist exclusively of Pick functions, since $W(x)$ is negative and $\Delta(x)\geqq 0$. Exactly one of these functions will satisfy the remaining equation.

Since we may now suppose that x_1 and x_l are distinct, we consider the two interpolation problems of order $2n$ obtained by omitting the highest order condition at each point. If $d<n$, each of these problems admits a unique solution, of order at most $n-1$, and since $2n-1$ conditions are enough to determine such a solution, those solutions are the same. It follows that there is a unique solution to the problem and this solution is a Pick function from the inductive hypothesis. It is obvious that its poles are outside the interpolation interval.

We finally must consider the case when $d=n$. Here we let $f_t(x)$ be the family of solutions of the problem of order $2n$ obtained by omitting the highest order condition at x_1. Select t so that $f_t(x)$ satisfies the omitted equation; the arguments of the previous chapter show that this is the unique solution to the interpolation problem. We show it to be a Pick function with its poles outside the interpolation interval as follows. Consider the problem of order $2n$ obtained by omitting the highest order condition at the point x_l which we may suppose to be different from x_1, and let $g_s(x)$ be the corresponding family of solutions. The particular $f_t(x)$ considered above is a member of this family. The eventual residue $r(x)$ associated with the system $f_t(x)$ is given by

$\Delta(x)\,W(x)$ where $W(x)$ is negative. Similarly, the eventual residue $\hat{r}(x)$ associated with the system $g_s(x)$ is of the form $\hat{\Delta}(x)\,\hat{W}(x)$ where $\hat{W}(x)$ is also negative. Now the polynomials $\Delta(x)$ and $\hat{\Delta}(x)$ are related by the equation

$$(x-x_1)\,\Delta(x)=(x-x_l)\,\hat{\Delta}(x)$$

which clearly shows that these two polynomials have opposite sign in the interval (x_1, x_l) and therefore that $r(x)$ and $\hat{r}(x)$ also have opposite sign in that interval. It follows that the function $f_t(x)=g_s(x)$ cannot have a pole in that interval, since the corresponding residue would have to be both positive and negative. The poles of that function are therefore outside the closed interval $[x_1, x_l]$ and are associated with negative residues. It follows that the solution is a Pick function as the theorem asserts. This completes the proof.

It is instructive to return to the study of the interpolation when $N=2n$ is even and the determinant of order n is not 0. There is a family of solutions $f_t(x)$ where the exceptional solutions are distinct. We consider the solution which is exceptional for the point x_k. There is no loss of generality in supposing the problem so paramaterised that $T(x_k)=t_k=k$ so that $f_k(x)$ is the exceptional solution associated with x_k. For values of t close to k let x_t be the pole of $f_t(x)$ which is near x_k; all the other poles are outside the interpolation interval. Hence for t close to k but not equal to that value we have

$$f_t(z) = \frac{r(x_t)}{z-x_t} + h_t(z).$$

Let C be a circle in the complex plane which contains the interpolation points in its interior and such that $f_t(z)$ has no poles on or within C for t sufficiently close to k. Now $h_t(z)$ is analytic within C for t near k or $t=k$ and so for z inside that circle

$$\begin{aligned} h_t(z) &= f_t(z) - \frac{r(x_t)}{z-x_t} \\ &= \frac{1}{2\pi i}\int_C [f_t(\zeta) - r(x_t)/(\zeta - x_t)]\,\frac{d\zeta}{\zeta - z} \end{aligned}$$

which reduces easily to

$$\frac{1}{2\pi i}\int_C f_t(\zeta)\,\frac{d\zeta}{\zeta - z}.$$

Since $f_k(z)$ is analytic within C we have $f_k(z)=h_k(z)$ inside that circle, while for t near k but unequal to it we have $f_t(x_k)=f_{k0}$. Passing to the limit as t approaches k and putting $z=x_k$ we have

$$h_k(x_k)=f_k(x_k)=f_{k0}-\lim_{t\to k}\frac{r(x_t)}{x_k-x_t}$$

$$=f_{k0}+\lim_{x\to x_k}\left[\frac{\Delta(x)-\Delta(x_k)}{x-x_k}\right]\frac{1}{W(x)}=f_{k0}+\Delta'(x_k)/W(x_k).$$

Let $\nu=\nu_k$; if this is not 0 then $\Delta'(x_k)=0$ and the exceptional solution fails to satisfy the condition of highest order at x_k, whence $f_k^{(\nu)}(x_k)$ is unequal to $f_{k\nu}$. We differentiate ν times to obtain

$$h_t^{(\nu)}(z)=f_t^{(\nu)}(z)-\frac{(-1)^\nu\nu!\,r(x_t)}{(z-x_t)^{\nu+1}}$$

and passing to the limit as before

$$f_k^{(\nu)}(x_k)=f_{k\nu}+\lim_{x\to x_k}\frac{r(x)\,\nu!}{(x-x_k)^{\nu+1}}.$$

We have $\Delta(x)=\Delta^{(\nu+1)}(x_k)(x-x_k)^{\nu+1}+H(x)(x-x_k)^{\nu+2}$ for some polynomial $H(x)$ and deduce finally that

$$f_k^{(\nu)}(x_k)=f_{k\nu}+\nu!\,\Delta^{(\nu+1)}(x_k)/W(x_k).$$

From the previous theorem we obtain another proof of Loewner's theorem, essentially the original proof given by Loewner himself.

Theorem. *A function $f(x)$ belonging to $P_n(a,b)$ for every integer $n>0$ is in $P(a,b)$.*

Proof. Select once and for all a Pick function $\psi(z)$ in $P(a,b)$ which is not rational. Let $\{x_i\}$ be a countable dense subset of the interval (a,b) and for each n let S_n consist of the first $2n+1$ points of that sequence. Let S' be a subset of S_n consisting of an even number of points and for positive values of ε consider the determinant of the extended Loewner matrix associated with S' and the function

$$f(x)+\varepsilon\psi(x).$$

Since the function belongs to all the classes $P_n(a,b)$ the associated determinant is nonnegative for $\varepsilon>0$ and is a non-constant polynomial in ε. It therefore follows that for sufficiently small values of ε the determinant is strictly positive, and because there are only finitely many such subsets S' of S_n to consider, all such associated determinants are positive for sufficiently small ε. We therefore select a sequence ε_n diminishing monotonically to 0 so that the functions

$$f(x)+\varepsilon_n\psi(x)$$

have positive associated determinants for subsets of S_n. In view of the previous theorem, then, there exists for each n a uniquely determined Pick function $\varphi_n(x)$ of degree exactly n which coincides with $f(x)+\varepsilon_n\psi(x)$ on the set S_n and which has its poles outside some interval containing S_n.

Choose c and d so that $a<c<d<b$; it is clear that for sufficiently large values of n the functions $\varphi_n(x)$ belong to $P(c,d)$ and from Lemma 4 of Chapter II we deduce the existence of a subsequence of that family converging in P to a limit $\varphi(x)$ in $P(c,d)$. The subsequence will converge uniformly on compact subsets of (c,d). However, we know that the sequence $\varphi_n(x)$ converges on a dense subset of (a,b) to $f(x)$. It follows that $f(x)$ coincides with $\varphi(x)$ on (c,d) and hence that $f(x)$ belongs to $P(c,d)$. Since the c and d were almost arbitrary in (a,b) it follows finally that $f(x)$ is in $P(a,b)$ as required.

Chapter XIV. The Interpolation of Monotone Matrix Functions

Let $f(x)$ be a sufficiently smooth function defined on an interval (a,b) of the real axis and let S be a set of N numbers in that interval. We write the elements of S in the form

$$x_1 \leqq x_2 \leqq \cdots \leqq x_N$$

and shall say that a function $\varphi(x)$ interpolates $f(x)$ at S if $\varphi(x)$ is a rational function of degree at most $N/2$ having the property that

$$[\lambda_1, \lambda_2, \ldots, \lambda_m]_f = [\lambda_1, \lambda_2, \ldots, \lambda_m]_\varphi$$

for every subset $\lambda_1, \lambda_2, \ldots, \lambda_m$ of S. Thus the function $\varphi(x)$ is a solution to the Cauchy interpolation problem associated with the points of S and the function $f(x)$.

In this chapter we study the interpolation of functions $f(x)$ belonging to the class $P_n(a,b)$ where we take $N=2n-1$. The interpolation function will then have to be of degree at most $n-1$ and if it exists is unique. However, one further convention must be made. The function $f(x)$ in $P_n(a,b)$ has regularity properties established in Chapter VII: the derivative of order $2n-3$ is continuous and even convex, and the derivative of order $2n-2$ exists outside a countable set and is a monotone function. If, for our interpolation, we take for S a set consisting of the point x_0 taken $2n-1$ times, the interpolation problem involves the derivative of order $2n-2$ and this may not exist at x_0. In this case we shall require that we take for $f^{(2n-2)}(x_0)$ any value in between the limits

$$\lim_{h \to 0} f^{(2n-2)}(x_0 - h) \quad \text{and} \quad \lim_{h \to 0} f^{(2n-2)}(x_0 + h).$$

With this convention our arguments will go through. However, it should be observed that the case when all the interpolation points coincide is almost always uninteresting, and we might equally well exclude it from consideration.

Theorem I. *If $f(x)$ belongs to $P_n(a,b)$ and S consists of $2n-1$ numbers from that interval there exists a function $\varphi(x)$ which interpolates $f(x)$ at S and belongs to the class $P(a,b)$.*

Proof. Choose x_0 in the interval (a,x_1) and x_{2n} in (x_{2n-1},b) and adjoin these numbers to S to obtain a set S' of cardinal $2n+1$. Let $\psi(z)$ be a function in $P(a,b)$ which is not rational, and for a small positive ε consider the interpolation problem associated with S' and the function

$$f(x)+\varepsilon\psi(x).$$

We claim that if ε is sufficiently small every associated determinant for this interpolation problem is positive. This follows from the fact that any given determinant is a polynomial in ε which is not constant and which is nonnegative for $\varepsilon>0$. The polynomial is therefore positive for sufficiently small positive ε, and only a finite number of such determinants need be considered. We can therefore invoke Theorem I of the previous chapter to deduce the existence of a rational function $\varphi_\varepsilon(z)$ in the Pick class of degree at most n which interpolates $f(x)+\varepsilon\psi(x)$ at S'. This function is regular in an interval containing S'. Let ε converge to 0 through some sequence; passing to a subsequence if need be we obtain a system $\varphi_\varepsilon(z)$ in the Pick class which converges to some limit $\varphi_0(z)$ also in the Pick class. Since the poles of $\varphi_\varepsilon(z)$ are outside the closed interval $[x_0,x_{2n}]$ it follows that for every subset $\lambda_1,\lambda_2,\ldots,\lambda_m$ of the set S we have

$$[\lambda_1,\lambda_2,\ldots,\lambda_m]_{\varphi_0}=[\lambda_1,\lambda_2,\ldots,\lambda_m]_f.$$

If d is the degree of $\varphi_0(z)$ then the interpolation problem associated with f and S has the property that every associated determinant of order $>d$ vanishes and every associated determinant of lower order is positive. The theorem of the previous chapter then guarantees the existence of a unique solution to that interpolation problem. We write that solution φ and know that it is a Pick function. It remains to show that this function belongs to $P(a,b)$.

Suppose d, the degree of $\varphi_0(z)$, is smaller than n. In this case $\varphi_0(z)$ and $\varphi(z)$ coincide, and since $\varphi_0(z)$ belongs to $P(x_0,x_{2n})$ so does $\varphi(z)$. On the other hand, if $d=n$, then for ε sufficiently small the poles of $\varphi_\varepsilon(z)$ are bounded away from the points x_0 and x_{2n} and it follows that $\varphi_0(x_0)=f(x_0)$ and that $\varphi_0(x_{2n})=f(x_{2n})$. Consider the interpolation problem associated with the set S, the point x_0 and the function $f(x)$. The Pick function $\varphi_0(z)$ is a solution to this problem, which is of order $2n$ and has a positive determinant. It is then clear that $\varphi(x)$ is the exceptional solution associated with the point x_0. This means that $\varphi(x)$ is regular in the interval (x_0,x_{2n-1}). Arguing in the same way with the set S and the point x_{2n} we infer that the function $\varphi(x)$ has no pole in the interval (x_1,x_{2n}). Thus, whatever the number d is, the interpolation function $\varphi(x)$ has no pole in (x_0,x_{2n}) and since those points were arbitrary it follows that that function belongs to $P(a,b)$. This completes the proof of the theorem.

In Chapter VII we noted that if a function $f(x)$ in $P_2(a,b)$ has a derivative which vanishes at some point then $f(x)$ is constant. This property has the following remarkable generalization to $P_n(a,b)$.

Theorem II. *If the interpolation function $\varphi(x)$ of the previous theorem is of degree at most $n-2$ then $f(x)$ and $\varphi(x)$ coincide throughout the interval (a,b).*

Proof. Choose a point x_0 in the interval (a,b) and substitute this in S for some number of that set to obtain a set S' of cardinal $2n-1$. The determinant of order $n-1$ associated with the $2n-2$ numbers that S and S' have in common must vanish, and the interpolation problem associated with that determinant has the unique solution $\varphi(x)$. It follows that $\varphi(x_0)=f(x_0)$ and since x_0 was arbitrary, the functions $f(x)$ and $\varphi(x)$ coincide in (a,b).

In view of what has just been shown, a function $f(x)$ in $P_n(a,b)$ either is a rational function in $P(a,b)$ of degree at most $n-2$ or every determinant of order $n-1$ which can be associated with that function by means of an arbitrary choice of $2n-2$ numbers in the interval is strictly positive.

In order to study the interpolation of $f(x)$ by $\varphi(x)$ we introduce a polynomial associated with the set S:

$$S(x)=\prod_{i=1}^{2n-1}(x-x_i)$$

and consider a point x_0 in the interval (a,b) which is not in the set S. We pass to the interpolation problem belonging to $f(x)$ and the $2n$ points obtained by adjoining x_0 to S. If $f(x)$ is not a rational Pick function of degree at most $n-2$ on the whole interval that interpolation problem surely has solutions, since all associated determinants of order $n-1$ or less are strictly positive, and the one determinant of order n is nonnegative. If that determinant vanishes, the interpolating Pick function is the unique solution to the problem and we have $f(x_0)=\varphi(x_0)$. When the determinant does not vanish $\varphi(x)$ is an exceptional solution to the problem and does not coincide with $f(x)$ at $x=x_0$. From an argument of the previous chapter we infer

$$f(x_0)-\varphi(x_0)=-\Delta'(x_0)/W(x_0)$$

and since $W(x_0)$ is negative and $\Delta'(x_0)=S(x_0)$ we may write finally

$$[f(x)-\varphi(x)]\,S(x)\geqq 0$$

for all x in the interval (a,b), since the choice of x_0 was arbitrary.

Suppose, next, that there exists a set of $2n$ numbers in the interval

$$x_0 \leqq x_1, \ldots, \leqq x_{2n-1}$$

such that the corresponding interpolation problem for $f(x)$ has an associated determinant of order n which vanishes. As we have seen, that problem then has a unique solution $\varphi(x)$ which is the interpolation function associated with the set S consisting of all the x's except x_0. On the other hand, $\varphi(x)$ is also the interpolation function belonging to the set $\hat{S}$ which consists of all the x's except x_{2n-1}. We therefore have the two inequalities

$$[f(x)-\varphi(x)]\,S(x) \geqq 0 \quad \text{and} \quad [f(x)-\varphi(x)]\,\hat{S}(x) \geqq 0\,.$$

Since $(x-x_0)S(x)=(x-x_{2n-1})\hat{S}(x)$ the polynomials $S(x)$ and $\hat{S}(x)$ have opposite signs in the interval (x_0, x_{2n-1}) and we infer that $f(x)$ and $\varphi(x)$ coincide in that interval.

It is now clear that if there exists a point x_0 in (a,b) such that $f(x_0)=\varphi(x_0)$ with x_0 not in S then $f(x)$ and $\varphi(x)$ coincide on an interval containing the numbers of S and x_0 and those functions are equal nowhere else. We have therefore established the following theorem.

Theorem III. *The interpolating function $\varphi(x)$ is related to $f(x)$ in one of the following ways:*

(i) *The functions coincide throughout (a,b) and are rational Pick functions of degree at most $n-2$.*

(ii) *The functions coincide throughout a subinterval of (a,b) where they are rational Pick functions of degree exactly $n-1$ and they are equal nowhere else.*

(iii) *The functions are equal only at points of S.*

In any event we have $[f(x)-\varphi(x)]\,S(x) \geqq 0$.

Our argument has also established the following useful result.

Theorem IV. *Let $f(x)$ belong to $P_n(a,b)$ and for some set of $2n$ numbers of that interval have an associated determinant which vanishes. Then $f(x)$ is rational, of degree at most $n-1$ on the smallest interval containing those numbers.*

Note that in the special case when all $2n$ numbers coincide the theorem says nothing.

We pass now to the proof of the important fact that membership in P_n is a local property. This completes the analysis of the class $P_n(a,b)$ given in Chapter VII.

Theorem V. *Let $a<c<b<d$ and suppose that $f(x)$ is defined on (a,d) and belongs to $P_n(a,b)$ and also to $P_n(c,d)$. Then the function belongs to $P_n(a,d)$.*

Proof. We first observe that it is enough to prove the theorem for smooth functions, since $f(x)$ is surely the pointwise limit of its smooth regularizations and $P_n(a,d)$ is closed under pointwise convergence. For the same reason, we may confine ourselves to the consideration of functions $g(x)$ of the form

$$g(x)=f(x)+\varepsilon\,\Psi(x)$$

where ε is small and positive and $\Psi(x)$ is a function in $P(a,d)$ which is not rational. We will evidently have

$$M_k(x,g)=M_k(x,f)+\varepsilon\,M_k(x,\Psi)$$

for all $k\leqq n$ and this equation in positive matrices makes it plain that $\det M_k(x,g)$ is strictly positive throughout the interval (a,d). It will follow that $g(x)$ cannot be rational, of degree at most $n-1$ over any subinterval of (a,d). We will show that the hypothesis that $g(x)$ is not in $P_n(a,d)$ contradicts this fact. For this purpose we first show that every associated determinant of order $k\leqq n-1$ is strictly positive. For if such a determinant vanished, it must be associated with a set of $2k$ points not all of which are equal, since $\det M_k(x,g)>0$ everywhere. From the previous theorem it follows that there exists a subinterval upon which $g(x)$ appears as a rational function of degree at most $n-1$, a contradiction.

Because the function is not in $P_n(a,d)$ there exists a set of $2n$ distinct numbers in the interval

$$\xi_1'<\eta_1'<\xi_2'<\cdots<\eta_n'$$

such that the corresponding Loewner determinant is negative. Evidently $\xi_1'\leqq c$ and $\eta_n'\geqq b$. Select a' in (a,ξ_1') and define b' as the supremum of all y such that $g(x)$ belongs to $P_n(a',y)$. It is easy to see that $b\leqq b'\leqq\eta_n'$ and that $g(x)$ belongs to $P_n(a',b')$. Because $g(x)$ is clearly not in $P_n(a',b'+\varepsilon)$ for $\varepsilon>0$ we can select a system of $2n$ distinct numbers

$$\xi_1(\varepsilon)<\eta_1(\varepsilon)<\xi_2(\varepsilon)<\cdots<\eta_n(\varepsilon)$$

in the interval $(a',b'+\varepsilon)$ such that the corresponding Loewner determinant is negative. Evidently $\xi_1(\varepsilon)\leqq c$ and $\eta_n(\varepsilon)\geqq b'$ and as ε converges to 0 through an appropriate sequence we obtain a set S_0^* of $2n$ numbers

$$\xi_1\leqq\eta_1\leqq\xi_2\leqq\cdots\leqq\eta_n$$

corresponding to an extended Loewner matrix with a non-positive determinant. Clearly $\eta_n=b'$ and $a'\leqq\xi_1\leqq c$. Here it is necessary to consider the determinant of the extended Loewner matrix because we do not know that the points ξ_i and η_j are all distinct. The determinant in question is in fact 0 since the function is in $P_n(a',b')$ and the points of

S_0^* are in the closed interval $[a', b']$; the determinant depends continuously on those points since the function is smooth, and from this it follows that the determinant is non-negative. Since the determinant also depends continuously on ε it is non-positive, hence equal to 0.

Let S_ε^* denote the set S_0^* translated ε units to the left where ε is positive and small. This set lies in the interval $(a' - \varepsilon, b')$ and $g(x)$ belongs to $P_n(a' - \varepsilon, b')$. Let $\varphi_\varepsilon(x)$ and $\hat{\varphi}_\varepsilon(x)$ be interpolating functions for $g(x)$ and the $2n-1$ points obtained from S_ε^* by omitting the number $\xi_1 - \varepsilon$ and the number $\eta_n - \varepsilon$ respectively. These interpolation functions correspond to polynomials $S_\varepsilon(x)$ and $\hat{S}_\varepsilon(x)$ with opposite signs in the interval $(\xi_1 - \varepsilon, b' - \varepsilon)$ and it follows from Theorem III that over this interval the values of $g(x)$ are in between the values of the interpolating functions. As ε approaches 0 these interpolating functions converge to a common limit, a rational function of degree $n-1$, the unique solution to the Cauchy interpolation problem corresponding to S_0^*. Evidently $g(x)$ coincides with this function on the interval (ξ_1, η_n) and this contradiction completes the proof.

Chapter XV. Almost Positive Matrices

A symmetric matrix $A=a_{ij}$ is said to be *almost positive* if the corresponding quadratic form satisfies the inequality

$$\sum\sum a_{ij} z_i \bar{z}_j \geqq 0$$

whenever $\sum z_i=0$. It is clear that a positive matrix is an almost positive matrix and that the class of almost positive matrices forms a convex cone. It is easy to see that any matrix A of the form

$$a_{ij}=\alpha_i+\bar{\alpha}_j$$

is an almost positive matrix, and therefore any matrix with

$$a_{ij}=b_{ij}+\alpha_i+\bar{\alpha}_j$$

where $B=b_{ij}$ is a positive matrix, is also almost positive. We first show that every almost positive matrix is of this form.

Lemma 1. *Let A be an almost positive matrix of order n; then the matrix B defined by*

$$b_{ij}=a_{ij}-a_{in}-a_{nj}+a_{nn}$$

is a positive matrix.

Proof. First note that B is certainly almost positive, since it differs from A by a matrix of the form $\alpha_i+\bar{\alpha}_j$ and hence

$$\sum\sum b_{ij} z_i \bar{z}_j \geqq 0$$

provided $\sum z_i=0$. However, the final row and final column of the matrix B consist entirely of zeros, and so the quadratic form is independent of the value z_n. We are therefore at liberty to set $z_n=-\sum_{i=1}^{n-1} z_i$ and infer that B is a positive matrix.

Our argument shows that the symmetric matrix of order 2

$$\begin{bmatrix} a & b \\ \bar{b} & c \end{bmatrix}$$

is an almost positive matrix if and only if $\mathrm{Re}[b] \leqq (a+c)/2$.

Lemma 2. *Let A be an almost positive matrix of order n; then its Schur exponential E defined by $e_{ij}=\exp a_{ij}$ is a positive matrix.*

Proof. Let $\alpha_i = a_{in} - (a_{nn}/2)$ and $\beta_i = \exp \alpha_i$. We have

$$a_{ij} = b_{ij} + \alpha_i + \overline{\alpha}_j$$

where b_{ij} is a positive matrix. Accordingly

$$e_{ij} = \beta_i \overline{\beta}_j \exp b_{ij}.$$

Thus E is the Schur product of the positive matrix $\beta_i \overline{\beta}_j$ and the positive matrix $\exp b_{ij}$ and is therefore a positive matrix itself.

As in the case of our study of reproducing kernels we extend the definition of almost positive matrices to general sets. Let S be an arbitrary set and $A(x,y)$ a function defined on $S \times S$ which is Hermitian symmetric. Such a function is an almost positive matrix if and only if for every finite subset x_i of S and equally many complex numbers z_i for which $\sum z_i = 0$ we have

$$\sum \sum A(x_i, x_j) z_i \overline{z}_j \geqq 0.$$

It is evident that any function of the form

$$A(x,y) = F(x) + \overline{F(y)}$$

is an almost positive matrix, and that any positive matrix $K(x,y)$ is an almost positive matrix. The family of almost positive matrices forms a convex cone, and Lemma 2 makes it clear that if $A(x,y)$ is almost positive then $K(x,y) = \exp A(x,y)$ is a positive matrix. Similarly, from Lemma 1 we deduce that for any z in S the function

$$K(x,y) = A(x,y) - A(z,y) - A(x,z) + A(z,z)$$

is a positive matrix on $S \times S$.

Another definition is useful. A positive matrix $K(x,y)$ on $S \times S$ which takes only positive values is infinitely divisible if and only if for every $s > 0$ the function $K^s(x,y)$ is a positive matrix.

The important theorem is the following.

Theorem I. *A function $K(x,y)$ which takes only positive values is an infinitely divisible positive matrix if and only if $A(x,y) = \log K(x,y)$ is almost positive.*

Proof. We first suppose that $K(x,y)$ is infinitely divisible and pick a finite set x_i in S and equally many complex numbers z_i so that $\sum z_i = 0$. The quantity

$$\sum \sum K^s(x_i, x_j) z_i \overline{z}_j$$

is positive and unchanged if we subtract off $\sum\sum z_i \bar{z}_j$. Dividing then by the positive s we find

$$\sum\sum \frac{K^s(x_i, x_j) - 1}{s} z_i \bar{z}_j \geqq 0$$

and passing to the limit as s approaches 0 we get finally

$$\sum\sum \log K(x_i, x_j) z_i \bar{z}_j \geqq 0$$

and therefore infer that $\log K(x,y)$ is an almost positive matrix. On the other hand, if the logarithm is almost positive, so also is $s \log K(x,y)$ for any $s>0$, and the exponential of this function is a positive matrix in view of Lemma 2. This means that $K^s(x,y)$ is a positive matrix, as desired.

Theorem II. *Let S be a metric space with metric $\rho(x,y)$; then S is isometric to a subset of Hilbert space if and only if the function $A(x,y) = -\rho(x,y)^2$ is an almost positive matrix.*

Proof. First suppose that S is isometric to a subset of Hilbert space and that the isometry is given by the mapping

$$x \to K_x .$$

We will then have a corresponding kernel function or positive matrix $K(x,y)$ as in Chapter X and

$$\rho(x,y)^2 = \|K_x - K_y\|^2 = K(x,x) + K(y,y) - 2\,\mathrm{Re}[K(x,y)] .$$

Since the real part of a positive matrix is again a positive matrix this makes

$$A(x,y) = -\rho(x,y)^2 = B(x,y) + F(x) + \overline{F(y)}$$

where $B(x,y) = 2\,\mathrm{Re}[K(x,y)]$ is a positive matrix and $F(x) = -K(x,x)$. Thus $A(x,y)$ appears as the sum of a positive matrix and an almost positive one and so is almost positive itself. Suppose next that $A(x,y) = -\rho(x,y)^2$ is an almost positive matrix and fix an element z in S to form

$$K(x,y) = A(x,y) - A(x,z) - A(z,y) + A(z,z) .$$

This is a positive matrix and so is $H(x,y) = \frac{1}{2} K(x,y)$. This gives rise to a mapping of S into Hilbert space

$$x \to H_x$$

and

$$\begin{aligned} \|H_x - H_y\|^2 &= H(x,x) + H(y,y) - 2H(x,y) \\ &= \tfrac{1}{2} K(x,x) + \tfrac{1}{2} K(y,y) - K(x,y) \\ &= -A(x,y) = \rho(x,y)^2 . \end{aligned}$$

Thus the mapping is an isometry and the proof is complete.

It is now convenient to introduce classes of functions corresponding to the classes $P_n(a,b)$ studied earlier in this book, but now associated with almost positive Pick matrices.

Let (a,b) be an interval of the real axis and n a positive integer. The class $Q_n(a,b)$ is defined as the class of all real C^1 functions $f(x)$ on (a,b) such that the corresponding Pick matrix

$$K_{ij} = [\lambda_i, \lambda_j]_f$$

is an almost positive matrix for every choice of the n distinct points λ_i in the interval. Evidently $Q_n(a,b)$ contains $P_n(a,b)$. The class is not defined for $n=1$. It is immediate that $Q_n(a,b)$ is a convex cone and that $Q_{n+1}(a,b)$ is contained in $Q_n(a,b)$. We should also note that if a sequence $f_k(x)$ in $Q_n(a,b)$ converges pointwise to a limit $f_0(x)$ in C^1 in such a way that the derivatives converge pointwise to the derivative of $f_0(x)$ then $f_0(x)$ itself is also in $Q_n(a,b)$. This is a consequence of the obvious fact that the Pick matrices for $f_k(x)$ and a given fixed set λ_i converge to the corresponding Pick matrix for $f_0(x)$.

We first study the class $Q_2(a,b)$.

Lemma 3. *A real C^1 function $f(x)$ belongs to the class $Q_2(a,b)$ if and only if its derivative $f'(x)$ is convex on the interval (a,b).*

Proof. The function $f(x)$ belongs to $Q_2(a,b)$ if and only if for every choice of the distinct points x and y of that interval the matrix

$$\begin{bmatrix} [x,x] & [x,y] \\ [x,y] & [y,y] \end{bmatrix}$$

is an almost positive matrix. Here we have omitted the subscript of the divided difference. But this matrix is almost positive if and only if

$$2[x,y] \leqq [x,x] + [y,y]$$

and this may be written in integral form as

$$\frac{1}{y-x} \int_x^y f'(t)\,dt \leqq \frac{f'(x)+f'(y)}{2}.$$

This inequality asserts that the average of the function $f'(x)$ over an interval (x,y) is no larger than its average over the end-points of that interval. Now if $f'(x)$ is convex it surely satisfies that inequality. On the other hand, we shall see that the inequality implies the convexity of $f'(x)$. Choose points x_1 and x_2 arbitrarily in (a,b) and choose constants A and B so that the auxiliary function

$$g(x) = f'(x) + Ax + B$$

satisfies the equations $g(x_1)=g(x_2)=0$. We shall show that $g(x)\leqq 0$ in (x_1,x_2), thus guaranteeing the convexity of $f'(x)$ since the points x_1 and x_2 were arbitrary. If there exists a point z in (x_1,x_2) such that $g(z)>0$ we let (y_1,y_2) be the largest (open) interval about z on which the continuous $g(x)$ is >0. Evidently $1/(y_2-y_1)\int_{y_1}^{y_2} g(x)dx$ is strictly positive, although $g(y_1)=g(y_2)=0$. But this contradicts the obvious fact that $g(x)$ satisfies the same integral inequality that $f'(x)$ does.

We can now consider the case for general values of n.

Lemma 4. *A real C^1 function $f(x)$ belongs to the class $Q_n(a,b)$ if and only if for every z in (a,b) the function $[x,z,z]_f$ belongs to $P_{n-1}(a,b)$.*

Proof. We will omit the subscripts on the divided differences in the sequel. The function $f(x)$ belongs to $Q_n(a,b)$ if and only if the Pick matrix $[\lambda_i,\lambda_j]$ is an almost positive matrix for any choice of n distinct values of λ_i in the interval. This matrix is almost positive if and only if

$$[\lambda_i,\lambda_j]-[\lambda_i,\lambda_n]-[\lambda_n,\lambda_j]+[\lambda_n,\lambda_n]$$

is a positive matrix. This matrix may be written

$$(\lambda_i-\lambda_n)(\lambda_j-\lambda_n)\,[\lambda_i,\lambda_j,\lambda_n,\lambda_n]$$

and since the final row and column consist only of zeros, this is a positive matrix if and only if the matrix of order $n-1$ obtained by omitting the final row and column is a positive matrix. Thus we permit the indices i and j to run only between 1 and $n-1$ in the sequel. The resulting matrix is a positive matrix if and only if its Schur product with the positive matrix $(\lambda_i-\lambda_n)^{-1}(\lambda_j-\lambda_n)^{-1}$ is a positive matrix. This means that the Pick matrix of the function $[x,\lambda_n,\lambda_n]$ associated with the $n-1$ points $\lambda_1,\lambda_2,\dots,\lambda_{n-1}$ is a positive matrix. The $n-1$ points are arbitrary points of the interval, subject only to the condition that they are distinct and different from λ_n.

Put $\lambda_n=z$ and suppose now that $f(x)$ belongs to $Q_n(a,b)$. The function $[x,z,z]$ is then surely in $P_{n-1}(a,z)$ as well as $P_{n-1}(z,b)$. It almost satisfies the criterion of Chapter VIII for membership in $P_{n-1}(a,b)$ except that the Pick matrix cannot yet be computed for a point $\lambda_i=z$. However, if $f(x)$ has a continuous third derivative the function $[x,x,z,z]$ varies continuously with the points of the interval. It follows that $[x,z,z]$ is in $P_{n-1}(a,b)$ when the third derivative of $f(x)$ is continuous. However, $f(x)$ can be approximated by its regularizations, smooth functions which belong to $Q_n(a+\varepsilon,b-\varepsilon)$ for small $\varepsilon>0$. Thus the corresponding functions $[x,z,z]_\varepsilon$ belonging to $P_{n-1}(a+\varepsilon,b-\varepsilon)$ converge pointwise to $[x,z,z]$ which is then also in that class.

On the other hand, if we suppose that the real C^1 function $f(x)$ has the property that for every z of the interval the function $[x,z,z]$ belongs to $P_{n-1}(a,b)$, we easily show that its Pick matrix must be almost positive. Given any n distinct points λ_i of the interval, the Pick matrix associated with those points is almost positive if and only if the Pick matrix of order $n-1$ of the function $[x,\lambda_n,\lambda_n]$ and the first $n-1$ points λ_i is a positive matrix. The positivity of this matrix follows from the fact that $[x,\lambda_n,\lambda_n]$ is in the class $P_{n-1}(a,b)$. The proof is complete.

Lemma 5. *Let $f(x)$ be a real C^1 function on the interval (a,b) and z a point of the interval with the property that $[x,z,z]$ belongs to $P_n(a,b)$. Then $f(x)$ is in $Q_n(a,b)$.*

Proof. Choose first n distinct points of the interval λ_i so that none of them equals z and adjoin z to the set to obtain $n+1$ distinct points. The Pick matrix of order $n+1$ computed for $f(x)$ and this set is almost positive if and only if the Pick matrix of order n computed for $[x,z,z]$ and the λ_i is a positive matrix. Our hypothesis that $[x,z,z]$ is in $P_n(a,b)$ guarantees that this is the case. Since a submatrix of an almost positive matrix is almost positive, it is clear that every Pick matrix computed for $f(x)$ and n distinct points of the interval is almost positive, and hence $f(x)$ is in $Q_n(a,b)$.

Let $Q(a,b)$ denote the intersection of all the classes $Q_n(a,b)$. Thus $Q(a,b)$ consists of all real C^1 functions on the interval such that the corresponding Pick matrix

$$\begin{aligned} K(x,y) &= \frac{f(x)-f(y)}{x-y} \qquad \text{for } x \neq y \\ &= f'(x) \qquad\qquad \text{for } x=y \end{aligned}$$

is an almost positive matrix. We would expect that there would be an analogue to Loewner's Theorem in this case which would make the functions in $Q(a,b)$ analytic. That this is in fact the case is shown by the following theorem.

Theorem III. *A real C^1 function $f(x)$ on the interval (a,b) belongs to $Q(a,b)$ if and only if for at least one point z of the interval (and therefore for all such points) the function $[x,z,z]_f$ is in $P(a,b)$.*

Proof. The assertions of the theorem are immediate consequences of the previous two lemmas and Loewner's theorem. We may therefore fix a point z of the interval and write

$$f(x)=f(z)+f'(z)(x-z)+(x-z)^2\,\varphi(x)$$

where $\varphi(x)$ belongs to $P(a,b)$.

Chapter XVI. The Analytic Continuation of Bergman Kernels

This chapter is devoted to a theorem of FitzGerald concerning the sesqui-analytic continuation of Bergman kernel functions. First, however, it is necessary to consider a certain reproducing kernel space.

Let $\mathscr{D}$ be an open disk in the plane and dA_z the usual two-dimensional Lebesgue measure, i.e., $dA_z = dx\,dy$. By $\mathscr{H}(\mathscr{D})$ we denote the linear space of all functions $f(z)$ analytic in $\mathscr{D}$ for which the integral

$$\|f\|^2 = \int_{\mathscr{D}} |f(z)|^2\, dA_z$$

is finite. This integral defines a quadratic norm on $\mathscr{H}(\mathscr{D})$.

Lemma 1. *$\mathscr{H}(\mathscr{D})$ is a Hilbert space with a reproducing kernel.*

Proof. We have only to show that the space is complete and that the value of a function at a fixed point z is a continuous linear functional on $\mathscr{H}(\mathscr{D})$. We show that latter property first.

Let z be fixed in $\mathscr{D}$ and r the distance from z to the boundary of the disk. Let $U_z(w)$ be defined in $\mathscr{D}$ by

$$\begin{aligned} U_z(w) &= \frac{1}{\pi r^2} \quad \text{for } |w-z|<r \\ &= 0 \qquad \text{otherwise.} \end{aligned}$$

If $f(w)$ belongs to $\mathscr{H}(\mathscr{D})$ it is analytic and therefore harmonic in the disk and so satisfies the mean value property for harmonic functions. We therefore have

$$f(z) = \frac{1}{\pi r^2} \int_{|w-z|<r} f(w)\, dA_w = (f, U_z)$$

where the inner product is taken in $L^2(\mathscr{D})$. From the Schwarz inequality we infer that

$$|f(z)| \leqq C_z \|f\|$$

where C_z is the reciprocal of $\sqrt{\pi}\, r$. Thus the valuation at z is a continuous linear functional on the space.

It is now easy to see that $\mathscr{H}(\mathscr{D})$ is complete, since a Cauchy sequence $f_k(z)$ is evidently a family of analytic functions, uniformly bounded on compact subsets of the disk and converging pointwise on the disk. Such a sequence converges uniformly on compact subsets of the disk to an analytic function $f_0(z)$. Since the sequence evidently converges in $L^2(\mathscr{D})$ it is clear that $f_0(z)$ is integrable square and that $\|f_k - f_0\|$ converges to 0.

Let us emphasize that the space $\mathscr{H}(\mathscr{D})$ is not a subspace of $L^2(\mathscr{D})$. The elements of $L^2(\mathscr{D})$, properly speaking, are equivalence classes of functions, two functions being equivalent when they coincide almost everywhere. On the other hand the elements of $\mathscr{H}(\mathscr{D})$ are functions and not equivalence classes. Let $A^2(\mathscr{D})$ be the subspace of $L^2(\mathscr{D})$ consisting of those equivalence classes which contain an analytic function—i.e., an element of $\mathscr{H}(\mathscr{D})$. We can then obtain an isometry between $A^2(\mathscr{D})$ and $\mathscr{H}(\mathscr{D})$ by making $f(z)$ in $\mathscr{H}(\mathscr{D})$ correspond to the equivalence class in $A^2(\mathscr{D})$ containing $f(z)$.

We should also note that the Bergman kernel function was first brought into mathematics precisely for the study of the space $\mathscr{H}(\mathscr{D})$, whose kernel $B(z,w)$ reduces to

$$B(z,w) = \frac{1}{\pi(1 - z\overline{w})^2}$$

when $\mathscr{D}$ is the disk $|z| < 1$.

Let I be a compact line segment in $\mathscr{D}$ and $L^2(I)$ the usual L^2-space over I relative to the usual linear Lebesgue measure on the segment. The functions of $\mathscr{H}(\mathscr{D})$, when restricted to I, are continuous and bounded there and hence are square integrable over I. There is therefore a restriction mapping R from $\mathscr{H}(\mathscr{D})$ to $L^2(I)$ which carries $f(z)$ in $\mathscr{H}(\mathscr{D})$ to the equivalence class in $L^2(I)$ of its restriction to I. This mapping is obviously a linear transformation from the one Hilbert space to the other, and we now show that it is continuous. If $f(z)$ belongs to $\mathscr{H}(\mathscr{D})$ and if $B(z,w)$ is the kernel function of that space then

$$f(z) = (f, B_z)$$

and so for x in I

$$|f(x)|^2 = |(f, B_x)|^2 \leqq \|f\|^2 B(x,x).$$

Accordingly

$$\|Rf\|^2 = \int_I |f(x)|^2 dx \leqq \|f\|^2 C^2 \quad \text{where} \quad C^2 = \int_I B(x,x)dx.$$

The continuous mapping R from $\mathscr{H}(\mathscr{D})$ to $L^2(I)$ has an adjoint R^* from $L^2(I)$ to $\mathscr{H}(\mathscr{D})$ defined by the equation

$$(Rf, u) = (f, R^* u)$$

valid for all f in $\mathscr{H}(\mathscr{D})$ and all u in $L^2(I)$. The operator R^* is continuous; however, for us the important fact is that the range of R^* is dense in $\mathscr{H}(\mathscr{D})$. This we show by supposing that $f(z)$ in $\mathscr{H}(\mathscr{D})$ is orthogonal to the range of R^*, whence $(f, R^* u) = (Rf, u) = 0$ for all u in $L^2(I)$. It follows that the restriction of $f(z)$ to I vanishes almost everywhere on the segment, hence that $f(z)$ vanishes identically in $\mathscr{D}$.

Lemma 2. *Let $K(z,w)$ be a sesqui-analytic function on $\mathscr{D} \times \mathscr{D}$ which is a positive matrix on $I \times I$ for some closed segment I contained in $\mathscr{D}$. Then $K(z,w)$ is a positive matrix throughout $\mathscr{D} \times \mathscr{D}$.*

Proof. We may suppose that $K(z,w)$ is bounded in $\mathscr{D} \times \mathscr{D}$, since otherwise we pass to a slightly smaller disk where the function is bounded and prove the lemma there, and the statement of the lemma will follow for $\mathscr{D}$ by continuity.

The continuous and bounded kernel $K(z,w)$ now defines an integral operator on the space $L^2(\mathscr{D})$ as follows:

$$(Kg)(z) = \int_{\mathscr{D}} K(z,w) g(w) \, dA_w = (g, K_z)$$

where $K_z(w) = \overline{K(z,w)}$ is an analytic function of w in $\mathscr{D}$. It is easy to verify that

$$\|Kg\| \leqq MA \|g\|$$

where A is the area of $\mathscr{D}$ and M a bound for the kernel. Thus the integral operator K is continuous. We should note also that since the measure $g(w)\,dA_w$ has finite total mass the function $(Kg)(z)$ is an analytic function of z and hence that it belongs to $A^2(\mathscr{D})$, or, indeed, to $\mathscr{H}(\mathscr{D})$. If g in $L^2(\mathscr{D})$ is orthogonal to $A^2(\mathscr{D})$, then

$$(Kg)(z) = (g, K_z) = 0$$

since K_z is in $A^2(\mathscr{D})$.

Suppose that z_i is a finite set of points in $\mathscr{D}$ and α_i a set of equally many complex numbers. For each i let $u_i(z)$ be a nonnegative, real continuous function on $\mathscr{D}$ which vanishes outside some small neighborhood of z_i and so chosen that $\int_{\mathscr{D}} u_i(z) \, dA_z = 1$. Put $g = \sum \alpha_i u_i$ to obtain a function in $L^2(\mathscr{D})$ and form

$$(Kg, g) = \sum \sum \alpha_i \overline{\alpha}_j \int_{\mathscr{D}} \int_{\mathscr{D}} K(z,w) \, u_i(w) \overline{u_j(z)} \, dA_z \, dA_w \, .$$

If the supports of the functions $u_i(z)$ are sufficiently small this sum is approximately

$$\sum \sum \overline{K(z_i, z_j)} \alpha_i \overline{\alpha}_j$$

and it is now clear that the sesqui-analytic $K(z,w)$ is a positive matrix if the corresponding integral operator is a positive operator, that is, if $(Kg,g)\geqq 0$ for all g in $L^2(\mathscr{D})$.

Let P be the projection in $L^2(\mathscr{D})$ onto the subspace $A^2(\mathscr{D})$. We have already seen that the range of K is contained in the range of P and that K vanishes on the null space of P. Accordingly $K=PKP$ and it is enough for us to verify that $(Kg,g)\geqq 0$ for all g in $A^2(\mathscr{D})$. If we now identify the elements of $A^2(\mathscr{D})$ with functions in $\mathscr{H}(\mathscr{D})$, regarding K as an integral operator on that space, we have only to show that $(Kg,g)\geqq 0$ for g in a dense subspace of $\mathscr{H}(\mathscr{D})$ since the integral operator is continuous. Such a dense subspace is the range of R^*. We therefore compute (KR^*u,R^*u) for some fixed u in $L^2(I)$.

$$(KR^*u)(z)=(R^*u,K_z)=(u,RK_z)=\int_I u(x)\overline{K(z,x)}\,dx\,.$$

Hence

$$\begin{aligned}(KR^*u,R^*u)&=(RKR^*u,u)=\int_I\int_I u(x)\overline{K(y,x)}\,\overline{u(y)}\,dx\,dy\\&=\int_I\int_I K(x,y)u(x)\overline{u(y)}\,dx\,dy\end{aligned}$$

a quantity which is obviously real and nonnegative since $K(x,y)$ is a positive matrix on $I\times I$. This completes the proof of the lemma.

Lemma 3. *Let $\mathscr{D}_1$ and $\mathscr{D}_2$ be two overlapping open disks in the complex plane and let $K(z,w)$ be a function sesqui-analytic in $\mathscr{D}_1\times\mathscr{D}_1$ as well as in $\mathscr{D}_2\times\mathscr{D}_2$. If $K(z,w)$ is a positive matrix in $\mathscr{D}_1\times\mathscr{D}_1$ then it also is a positive matrix in $\mathscr{D}_2\times\mathscr{D}_2$ and every function in the reproducing kernel space over $\mathscr{D}_1$ admits an analytic continuation to a function in the reproducing kernel space over $\mathscr{D}_2$. Moreover, this continuation is an isometry between the two spaces.*

Proof. Select an interval I in the intersection $\mathscr{D}_1\cap\mathscr{D}_2$. Since $K(z,w)$ is a positive matrix on $\mathscr{D}_1\times\mathscr{D}_1$, it is surely a positive matrix on $I\times I$, and from the previous lemma we deduce that $K(z,w)$ is a positive matrix on $\mathscr{D}_2\times\mathscr{D}_2$.

Consider the family of functions $K(z,z_n)$ where z_n is a countable dense subset of the intersection $\mathscr{D}_1\cap\mathscr{D}_2$. These functions are analytic in the union $\mathscr{D}_1\cup\mathscr{D}_2$ and belong to both reproducing kernel spaces. The inner product between $K(z,z_n)$ and $K(z,z_m)$ in either reproducing kernel space is given by $K(z_m,z_n)$ and so is the same in either space. Thus we can apply the Gram-Schmidt process to these functions to obtain a system $u_n(z)$ analytic in the union $\mathscr{D}_1\cup\mathscr{D}_2$ and which is a complete orthonormal set in each reproducing kernel space.

Let $f(z)$ be a function in the reproducing kernel space over $\mathscr{D}_1$; it has a unique expansion

$$f(z) = \sum a_n u_n(z)$$

valid for z in $\mathscr{D}_1$ with coefficients which are summable square. This expansion also makes sense over $\mathscr{D}_2$ and clearly coincides with $f(z)$ on the intersection of the two disks. It follows that $f(z)$ admits an analytic continuation to $\mathscr{D}_2$ and the continuation appears as a function in the reproducing kernel space over $\mathscr{D}_2$ with norm $\sqrt{\sum |a_n|^2}$, and this is exactly the norm of $f(z)$ in the first space. Thus the continuation is an isometry. Since the two reproducing kernel spaces are on an equal footing in this argument, every function in the second space is continuable to one in the first, and the continuation is an isometry between the two spaces. The proof is complete.

We can now state and prove FitzGerald's theorem.

Theorem. *Let $\mathscr{D}$ be an open, simply connected subset of the complex plane containing a line segment I. Let G be an open connected subset of $\mathscr{D} \times \mathscr{D}$ containing $I \times I$ and the diagonal, i.e., every point of the form (z,z) with z in $\mathscr{D}$. If $K(z,w)$ is a sesqui-analytic function defined in G which is a positive matrix on $I \times I$, then $K(z,w)$ has a unique extension to a Bergman kernel function defined throughout $\mathscr{D} \times \mathscr{D}$.*

Proof. For each z in $\mathscr{D}$ let $\mathscr{C}_z$ be the largest open disk with center z such that $\mathscr{C}_z \times \mathscr{C}_z$ is contained in G. Select a point x in I. For every w in $\mathscr{D}$ there exists a Jordan arc in $\mathscr{D}$ connecting w and x, and this compact set is covered by the union of a finite number of discs $\mathscr{C}_{z_i}$. Repeated application of the previous lemma then guarantees that $K(z,w)$ is a positive matrix over $\mathscr{C}_{z_i} \times \mathscr{C}_{z_i}$ for each i, and therefore that $K(z,w)$ is a positive matrix in $\mathscr{C}_w \times \mathscr{C}_w$. The function is therefore a positive matrix in a neighborhood of every point of the form (z,z). Let $u_n(z)$ be a complete orthonormal set in the reproducing kernel space over $\mathscr{C}_x$; again from the previous lemma it follows that the functions in this set admit analytic continuations along the arc to functions in the reproducing kernel space over C_w, and these continuations form a complete orthonormal set in the final space since the continuation is an isometry. We infer, then, that

$$K(z,z) = \sum |u_n(z)|^2 < \infty$$

for every z in $\mathscr{D}$. The continuations $u_n(z)$ are well-defined since $\mathscr{D}$ is simply connected. Since the sum above converges, the function

$$K_1(z,w) = \sum u_n(z) \overline{u_n(w)}$$

makes sense throughout $\mathscr{D} \times \mathscr{D}$ and evidently is a positive matrix there. We will presently show that $K_1(z,w)$ is sesqui-analytic; then, because it coincides with $K(z,w)$ on every set of the form $\mathscr{C}_z \times \mathscr{C}_z$ it evidently coincides with it in G. It is therefore a Bergman kernel function which is an extension of the given $K(z,w)$.

Let w be fixed in $\mathscr{D}$. The partial sums of the series defining $K_1(z,w)$ satisfy the inequality

$$\left|\sum_{n=1}^{N} u_n(z)\overline{u_n(w)}\right|^2 \leqq K(z,z)\,K(w,w)$$

and are therefore uniformly bounded on any compact subset of $\mathscr{D}$. These partial sums therefore form an equicontinuous family and an appropriate subsequence converges to an analytic limit, i.e., for fixed w, $K_1(z,w)$ is analytic in z. Since $K_1(z,w)=\overline{K_1(w,z)}$ the function is evidently anti-analytic in w, and hence $K_1(z,w)$ is sesqui-analytic in $\mathscr{D}\times\mathscr{D}$. This completes the proof.

Chapter XVII. The Loewner-FitzGerald Theorem

In this chapter we consider a function $\varphi(x)$ in $P(a,b)$ which corresponds to the Pick matrix $K(x,y)$. As we have seen, $K(x,y)$ is a positive matrix on $(a,b)\times(a,b)$ and indeed admits a sesqui-analytic continuation to a Bergman kernel function $K(z,w)$ defined on $\mathscr{D}\times\mathscr{D}$ where $\mathscr{D}$ is the union of the interval (a,b) and the open upper and lower half-planes. The remarkable theorem which follows gives a necessary and sufficient condition for $\varphi(z)$ to be univalent, that is to say, for that function to be a conformal mapping of $\mathscr{D}$ onto $\varphi(\mathscr{D})$.

Theorem. *If $\varphi(z)$ in $P(a,b)$ is not a constant, the following four assertions are equivalent.*

(i) $A(x,y)=\log K(x,y)$ *is an almost positive matrix on* $(a,b)\times(a,b)$.

(ii) $K(x,y)$ *is an infinitely divisible positive matrix on* $(a,b)\times(a,b)$.

(iii) $S(x,y)=K(x,y)^{\frac{1}{2}}$ *is a positive matrix on* $(a,b)\times(a,b)$ *where, of course, the positive square root is taken.*

(iv) $\varphi(z)$ *is univalent.*

Proof. The equivalence of (i) and (ii) has already been shown in Chapter XV, while the definition of infinite divisibility shows that (ii) implies (iii). We next show that (iv) follows from (iii).

For this purpose we consider the Bergman kernel $K(z,w)$ defined on $\mathscr{D}\times\mathscr{D}$ and note that in the neighborhood of any point where $K(z,w)$ is not zero there exist exactly two sesquianalytic determinations of the square root. In particular, we pass to the open subset of $\mathscr{D}\times\mathscr{D}$ defined by the inequality $\operatorname{Re}[K(z,w)]>0$ and take for G the connected component of that set which contains the diagonal, i.e., all points of the form (z,z) with z in $\mathscr{D}$. Let $S(z,w)$ be defined in G as that determination of the square root which takes values in the right half-plane: this is sesquianalytic in G and coincides with the positive square root on $(a,b)\times(a,b)$. By virtue of (iii) and the FitzGerald continuation theorem given in the previous chapter we infer that $S(z,w)$ admits a unique sesquianalytic continuation over the whole of $\mathscr{D}\times\mathscr{D}$ where it is obvious that the equation

$$S(z,w)^2=K(z,w)$$

is valid.

Suppose next that the function $\varphi(z)$ is not univalent: there will then exist two distinct points z_1 and z_2 in $\mathscr{D}$ for which $\varphi(z_1)=\varphi(z_2)$. The nonconstant $\varphi(z)$ is real only on the interval (a,b), so if z_1 belonged to that interval so would z_2. However, since $\varphi(x)$ is strictly increasing on that interval the equation $\varphi(z_1)=\varphi(z_2)$ could not hold. Thus both points are outside the real axis. They must lie in the same half-plane, since $\operatorname{Im}[\varphi(z)]$ is negative in the lower half-plane. In view of the reflection principle, then, we may assume that both points are in the upper half-plane. Let $w_1=\bar{z}_2$; now $z_1-\bar{w}_1$ is not 0 and

$$K(z_1,w_1)=\frac{\varphi(z_1)-\overline{\varphi(w_1)}}{z_1-\bar{w}_1}=0\,.$$

It is clear that $S(z_1,w_1)=0$ as well. Differentiating $K(z,w)$ with respect to z we find

$$\frac{\partial K(z,w)}{\partial z}=2S(z,w)\frac{\partial S(z,w)}{\partial z}=\frac{\varphi'(z)}{z-\bar{w}}-\frac{\varphi(z)-\overline{\varphi(w)}}{(z-\bar{w})^2}$$

and putting $z=z_1$ and $w=w_1$ this reduces to

$$0=\frac{\varphi'(z_1)}{z_1-\bar{w}_1}$$

and because the denominator is not 0 this becomes simply $\varphi'(z_1)=0$. It follows that $K(z,w)=0$ only at points where z is a zero of the derivative $\varphi'(z)$. From the symmetry of the kernel and the reflection principle, $\varphi(z)$ and $\varphi'(z)$ both being real on (a,b), it follows that w is also a zero of $\varphi'(z)$. We finally infer that $K(z,w)=0$ only for points z and w which are zeros of the analytic $\varphi'(z)$.

Since the zeros of $\varphi'(z)$ are isolated, we can find a disk $\mathscr{C}_1$ contained in $\mathscr{D}$ and having center z_1 and which contains no other zero of the function. Similarly, there will exist a disk $\mathscr{C}_2$ with center w_1 lying wholly in $\mathscr{D}$ and containing no other zero of $\varphi'(z)$. Thus the sesquianalytic $K(z,w)$, when considered in the product $\mathscr{C}_1\times\mathscr{C}_2$ will have only one zero, namely that at (z_1,w_1). It follows that the function $F(z,w)=K(z,\bar{w})$ is analytic in $\mathscr{C}_1\times\overline{\mathscr{C}}_2$ and has an isolated zero at the point $(z_1,\bar{w}_1)$. It is well known that this cannot happen, although, for the sake of completeness, we shall presently give the proof. This contradiction completes the proof that (iii) implies (iv).

Lemma. *An analytic function $F(z,w)$ of two complex variables cannot have an isolated zero.*

Proof. We suppose that $F(z,w)$ has an isolated zero and deduce a contradiction. There is no loss of generality in our supposing that the function is defined in $\mathscr{C}\times\mathscr{C}$ where $\mathscr{C}$ is a disk of radius 2 about the origin and that the unique zero is the point $(0, 0)$. The reciprocal $G(z, w) = F(z, w)^{-1}$ is then analytic in the open set obtained by deleting $(0, 0)$ from $\mathscr{C} \times \mathscr{C}$ but is not bounded near the deleted point. If the fixed w is not 0, $G(z, w)$ is analytic for z in $\mathscr{C}$ and so, for $|z| < 1$ we have

$$G(z, w) = \frac{1}{2\pi i} \int_0^{2\pi} \frac{G(e^{i\theta}, w)}{e^{i\theta} - z} i e^{i\theta} d\theta .$$

We argue similarly with $G(e^{i\theta}, w)$ to find that

$$G(z, w) = \frac{1}{4\pi^2} \int_0^{2\pi} \int_0^{2\pi} \frac{G(e^{i\theta}, e^{i\psi})}{(e^{i\theta} - z)(e^{i\psi} - w)} e^{i(\theta + \psi)} d\theta d\psi$$

provided $|z| < 1$ and $|w| < 1$. Evidently this function is bounded in a neighborhood of $(0, 0)$, since $G(e^{i\theta}, e^{i\psi})$ is continuous on the compact set over which the integration is taken. This is the desired contradiction.

It remains to show that (iv) implies (i). For this purpose we consider a finite set of points x_i in the interval (a, b) and an equal number of complex quantities z_i such that $\sum z_i = 0$. We then assign an entropy to the function $\varphi(z)$ as the following number:

$$E = \sum \sum \log K(x_i, x_j) z_i \bar{z}_j .$$

Note that the entropy is always real, since $\log K(x, y)$ is real and symmetric. We have to show that if $\varphi(z)$ is univalent in $P(a, b)$ then its entropy is nonnegative for any choice of the x_i and the numbers z_i.

Suppose next that $\varphi(z)$ is a linear fractional transformation in $P(a, b)$ of the form

$$\beta + \frac{m}{\lambda - z}$$

where λ is not in the interval (a, b). We will have

$$K(x, y) = \frac{m}{(\lambda - x)(\lambda - y)}$$

even when $x = y$, and therefore

$$\log K(x, y) = \log m - \log |\lambda - x| - \log |\lambda - y| .$$

It is therefore clear that

$$\sum \sum \log K(x_i, x_j) z_i \bar{z}_j = 0$$

if we have $\sum z_i = 0$. We can argue similarly with the linear fractional transformation $\varphi(z) = \alpha z + \beta$ to find that the linear fractional transformations in $P(a,b)$ all have zero entropy relative to any set of points x_i in the interval and complex coefficients z_i. Thus the theorem is proved for such special functions.

Again, suppose that the non-constant $\varphi(z)$ is in $P(a, b)$ and that $\psi(z)$ is in the Pick class and that the composed function $\psi\circ\varphi$ makes sense for x in (a, b). The Pick matrix of the composed function is

$$\frac{\psi(\varphi(x))-\psi(\varphi(y))}{x-y}=\frac{\psi(\varphi(x))-\psi(\varphi(y))}{\varphi(x)-\varphi(y)}\cdot\frac{\varphi(x)-\varphi(y)}{x-y}$$

and is therefore the Schur product of Pick matrices. Taking logarithms we find

$$\log\frac{\psi(\varphi(x))-\psi(\varphi(y))}{x-y}=\log\frac{\psi(\varphi(x))-\psi(\varphi(y))}{\varphi(x)-\varphi(y)}+\log\frac{\varphi(x)-\varphi(y)}{x-y}.$$

It follows immediately that the entropy of the composed function relative to the set x_i with coefficients z_i is the sum of the entropy of $\varphi(x)$ relative to that set of points and coefficients and the entropy of $\psi(x)$ relative to the points $x_i'=\varphi(x_i)$ and the same coefficients z_i.

This circumstance enables us to make certain simplifications in our problem. Given $\varphi(z)$ in $P(a, b)$ which is univalent and given the points x_i in the interval with coefficients z_i also given, we pass to the composed function $\psi_1\circ\varphi\circ\psi_2$ where ψ_1 and ψ_2 are linear fractional transformations in the Pick class with real coefficients. We let ψ_2 map an interval (a', b') containing the points $+1$ and -1 onto (a, b) in such a way that the points x_i are contained in the open interval $(\psi_2(-1), \psi_2(+1))$. Similarly we let ψ_1 be a linear fractional transformation with real coefficients so chosen that the composed function $\psi_1\circ\varphi\circ\psi_2$ leaves the points $-1, 0$ and $+1$ fixed. The entropy of $\varphi(z)$ relative to x_i and z_i is the same as the entropy of the composed function relative to the points $x_i'=\psi_2^{-1}(x_i)$ and the same coefficients z_i. It follows that in the proof of our theorem we need only consider a special semi-group of univalent Pick functions, namely those which are regular in some interval containing the points $-1, 0$ and $+1$ and which leave those points fixed.

We can now generate a substantial class of functions in that semi-group for which the theorem is true. Let $k(t)$ be a continuous function defined on the closed interval $[0, T]$ and taking values in the interval $|x|<1$. Define a by $1/a=\sup|k(t)|$; we will have $a>1$. Consider next the differential equation

$$\frac{dy}{dt}=\frac{y^3-y}{1-k(t)y}=\theta(y, t)$$

with the initial condition $y(0)=z$ where z is taken in the domain $\mathscr{D}$ consisting of the union of the interval $(-a, a)$ with the open upper and lower half-planes. Since θ is analytic for y in $\mathscr{D}$ and is continuous for t in the interval $[0, T]$ the well-known Picard theorem asserts the existence and uniqueness of the solution $y(t)$ and this solution depends analytically on the initial condition. We have therefore a family of analytic functions $y(t, z)$ and it is important for us to notice that these functions are in the Pick class. For suppose z in the upper half-plane and consider the trajectory $y(t, z)$ as t runs through the interval. We must show that the trajectory is always in the upper half-plane, i.e., that $y(t, z)$ never becomes real. If, now, for some value of $t_0>0$ the point $y(t_0, z)$ is on the real axis, and the curve $y(t, z)$ for t in the interval $(0, t_0)$ is such that $k(t)y(t, z)$ is bounded away from $+1$, then that curve lies wholly in a region where $\theta(y, t)$ is Lipschitzian in y and the usual uniqueness theorems hold. However, from the reflection principle it is clear that the curve $y(t,\bar{z})=\overline{y(t, z)}$ also meets the real axis at the same point contrary to the uniqueness theorem. It follows that the curve can only meet the real axis when the quantity $1-k(t)y(t, z)$ becomes very small. It is clear that the function $\theta(y, t)$ is rational in y for fixed t; it can therefore be put in the form

$$\theta(y, t)=A y^2+B y+C+\frac{m}{\lambda-y}$$

where $\lambda=\lambda(t)=1/k(t)$ and the coefficients A, B, C and m all are continuous functions of t. The number m is always positive, since

$$m=\lim_{y\to\lambda}(\lambda-y)\theta(y, t)=\lim_{y\to\lambda}(y^3-y)\lambda=\lambda^4-\lambda^2\geqq a^4-a^2>0\,.$$

It follows, therefore, that when $y(t, z)$ is close to $1/k(t)$, the derivative dy/dt has a very large and positive imaginary part, i.e., the point is moving rapidly away from the real axis and the curve does not cross that axis. Thus $y(t, z)$ is in the upper half-plane so long as z is. Not only are the functions $y(t, z)=\varphi_t(z)$ in the Pick class, they are also univalent in view of the uniqueness theorem for solutions of the differential equation. We can also see that $\varphi_t(-1)=-1$, that $\varphi_t(0)=0$ and that $\varphi_t(+1)=+1$.

From the representation given above for $\theta(y, t)$ it should be clear that for fixed t that function is the sum of a quadratic expression in y and a Pick function: its Pick matrix, therefore, considered over the interval $(-1, +1)$ is an almost positive matrix. This fact, as we shall see in a subsequent chapter, is closely related to the fact that the solutions of the differential equation are functions in the Pick class.

Choose a system of points x_i in the interval $|x|<1$ and select equally many complex coefficients z_i with $\sum z_i=0$. We write the entropy of $\varphi_t(z)$ relative to these points and coefficients:

$$E(t)=\sum\sum \log K_t(x_i, x_j)z_i\bar{z}_j$$

where $K_t(x, y)$ is of course the Pick matrix associated with $\varphi_t(z)$. It is easy to see that $E(t)$ is a differentiable function of t. We consider the derivative

$$\frac{d}{dt}\log K_t(x, y) = \frac{d}{dt}\log\frac{\varphi_t(x)-\varphi_t(y)}{x-y} = \frac{d}{dt}\log(\varphi_t(x)-\varphi_t(y))$$
$$= \frac{\theta(\varphi_t(x), t)-\theta(\varphi_t(y), t)}{\varphi_t(x)-\varphi_t(y)}$$

and we may therefore write

$$\frac{dE(t)}{dt} = \sum\sum \frac{\theta(\varphi_t(x_i), t)-\theta(\varphi_t(x_j), t)}{\varphi_t(x_i)-\varphi_t(x_j)} z_i \bar{z}_j$$

a quantity which is positive since the Pick matrix associated with $\theta(y, t)$ is almost positive for any fixed t. Since it is evident that $E(0)=0$, the entropy of the function $\varphi(z)=z$ being obviously 0, the entropy of the function $\varphi_t(z)$ is positive.

This virtually completes the proof of the theorem, since in the next chapter we shall show that a univalent function $\varphi(z)$ in the class $P(-a, a)$ where $a>1$ which leaves the points $-1, 0$ and 1 fixed can be approximated uniformly on the closed interval $[-1, 1]$ by functions of the form $\varphi_t(z)$ satisfying the slightly more complicated differential equation

$$\frac{dy}{dt} = (y^3 - y)\left[\frac{p(t)}{1-k(t)y} + q(t)\right] = \theta(y, t)$$

where the continuous $k(t)$ takes values in $(-1, 1)$ and the coefficients $p(t)$ and $q(t)$ are measurable, non-negative and satisfy $p(t)+q(t)=1$. The Pick matrix associated with $\theta(y, t)$ is almost positive, and so the solutions $\varphi_t(z)$ will have positive entropy. It follows that φ does too.

An important feature of the proof of the Loewner-FitzGerald theorem is the fact that the entropy of a univalent Pick function is increased, or at least not diminished, if it is composed with another univalent Pick function. This fact has a converse: if one univalent function always has a greater entropy than another, then the first can be obtained from the second by composition with a suitable Pick function. More precisely, we have the following theorem due to FitzGerald.

Theorem II. *Let $f(x)$ and $g(x)$ be univalent functions in $P(a, b)$ corresponding to Pick matrices $F(x, y)$ and $G(x, y)$ respectively. Suppose that*

$$A(x, y) = \log F(x, y) - \log G(x, y)$$

is an almost positive matrix; then there exists a univalent Pick function $\Phi(z)$ so that

$$f(z) = (\Phi \circ g)(z).$$

Proof. Let (α, β) be the interval upon which g maps (a, b). The function $f \circ g^{-1}$ is a real, monotone, C^1 function on (α, β) and has there the Pick matrix

$$\frac{(f \circ g^{-1})(g(x)) - (f \circ g^{-1})(g(y))}{g(x) - g(y)} \quad \text{for } x \neq y$$

and

$$\frac{f'(g^{-1}(g(x)))}{g'(x)} \quad \text{for } x = y .$$

It is obvious that the logarithm of this Pick matrix is $A(x,y)$ and hence that $f \circ g^{-1} = \Phi$ belongs to $P(\alpha, \beta)$ and is univalent. It follows then that $f = \Phi \circ g$ as asserted.

We can also obtain an extension of Theorem III, Chapter XV by invoking the Loewner-FitzGerald theorem.

Theorem III. *Let $f(x)$ be a real C^1 function which is strictly increasing on the interval (a, b) and which maps that interval onto (α, β). If $-f(x)$ has an almost positive Pick matrix then $f(x)$ is the inverse of a univalent function in $P(\alpha, \beta)$.*

Proof. We first invoke Theorem III of Chapter XV to see that $-f(x)$ is analytically continuable throughout the upper and lower half-planes and therefore that $f(x)$ itself is analytic in the same domain. The function f^{-1} is a real C^1 function on (α, β) and we write its Pick matrix in the form $F(f(x), f(y))$. Let $K(x, y)$ be the Pick matrix for $f(x)$ on (a, b). Since $f(x)$ was strictly increasing, both F and K take strictly positive values and these functions are related by the equation

$$F(f(x), f(y)) = \frac{f^{-1}(f(x)) - f^{-1}(f(y))}{f(x) - f(y)} = \frac{x - y}{f(x) - f(y)} = K(x, y)^{-1} .$$

Now write the integral representation of the Gamma function for positive s, putting $t = k\lambda$ to obtain

$$\Gamma(s) = \int_0^\infty e^{-t} t^{s-1} dt = k^s \int_0^\infty e^{-\lambda k} \lambda^{s-1} d\lambda$$

and therefore

$$k^{-s} = \frac{1}{\Gamma(s)} \int_0^\infty e^{-\lambda k} \lambda^{s-1} d\lambda .$$

It follows that

$$K(x, y)^{-s} = \frac{1}{\Gamma(s)} \int_0^\infty \exp[-\lambda K(x, y)] \lambda^{s-1} d\lambda .$$

By hypothesis the Pick matrix of $-f(x)$ is almost positive, and therefore $-\lambda K(x, y)$ is almost positive when $\lambda>0$. We see that the exponential is a positive matrix as is its integral relative to the positive measure $\lambda^{s-1} d\lambda/\Gamma(s)$. Since $K(x, y)>0$ throughout $(a, b)\times(a, b)$ the integral always converges. Hence $K(x, y)^{-s}=F(f(x), f(y))^s$ is a positive matrix for all $s>0$, and this means that f^{-1} belongs to $P(\alpha, \beta)$ and is univalent.

Chapter XVIII. Loewner's Differential Equation

In this chapter we study a class of conformal mappings belonging to P and show that a certain dense subset of this class consists of mappings obtained from a differential equation. This fact was needed for the complete proof of the Loewner-FitzGerald theorem of the previous chapter.

Let $\mathcal{S}$ be the class of all functions φ in P having the following properties:

(i) for some $a>1$ φ belongs to $P(-a, a)$

(ii) the function leaves the points $-1, 0$ and 1 fixed

(iii) φ is univalent.

We note that $\mathcal{S}$ is a semi-group under composition: if φ_1 and φ_2 are in $\mathcal{S}$ then so is their composition $\varphi_1 \circ \varphi_2$.

If for some function φ in $\mathcal{S}$ we form the Loewner matrix associated with the set

$$\xi_1 = -1, \qquad \eta_1 = 0, \qquad \xi_2 = 0, \qquad \eta_2 = 1$$

then from Theorem V of Chapter III the determinant

$$\det \begin{bmatrix} [-1,0] & [-1,1] \\ [0,0] & [0,1] \end{bmatrix} = \det \begin{bmatrix} 1 & 1 \\ \varphi'(0) & 1 \end{bmatrix} = 1 - \varphi'(0)$$

is positive and equals 0 only for the simple function $\varphi(z) = z$. Accordingly, since φ is not constant and belongs to $P(-1, 1)$

$$0 < \varphi'(0) \leqq 1$$

with equality only for the identity function.

It is important to notice that different functions in $\mathcal{S}$ map the upper half-plane onto different images, for if φ_1 and φ_2 were functions with the same image, the composition $\varphi_1^{-1} \circ \varphi_2$ would map the half-plane onto itself and would belong to $P(-a, a)$ for some $a>1$. This function would leave the points $-1, 0$ and 1 fixed. A consideration of the integral representation of a Pick function makes it clear that the only functions in P which map the half-plane onto itself are linear fractional transformations. Since the identity mapping is the only linear fractional transformation leaving three points fixed we infer that φ_1 and φ_2 are the same.

Let us also observe that every function in $\mathscr{S}$, and indeed, every function which we shall consider in this chapter, is continuable across the interval $(-1, 1)$ by reflection. We make the convention that the image of φ in $\mathscr{S}$ is the image of the upper half-plane under φ. Of course an interval of the form $(-a, a)$ with $a>1$ is a subset of the boundary of the image.

Lemma 1. *Let φ_n be a sequence in $\mathscr{S}$ such that the images Ω_n form an increasing sequence of sets. Then the sequence converges in P to a function φ_0 in $\mathscr{S}$ with image $\Omega = \bigcup_n \Omega_n$.*

Proof. Since the sequence belongs to $P(-1, 1)$ and is uniformly bounded on the closed interval $[-1, 1]$ it follows from Lemma 4 of Chapter II that there exists a subsequence converging in $P(-1, 1)$ to a limit $\varphi_0(z)$. If the sequence itself did not converge in P there would exist different subsequences converging to different limits. We shall show that the limit is unique, thereby assuring the convergence of the sequence. To show this, we first show that $\varphi_0(z)$ is univalent. We will then show that its image is exactly what the lemma asserts it to be, and this will make the function unique. It is obvious that $\varphi_0(z)$ leaves the points -1, 0 and 1 fixed and so is an element of $\mathscr{S}$.

Since φ_0 leaves the points $-1, 0$ and 1 fixed it cannot be a constant. Suppose that there exist two different numbers z_1 and z_2 in the domain of φ_0 such that $\varphi_0(z_1) = \varphi_0(z_2) = \lambda$. Let C be a smooth curve surrounding these points and wholly contained in the domain of φ_0 upon which the non-constant φ_0 does not take the value λ. The integral

$$W(\lambda, \varphi_0, C) = \frac{1}{2\pi i} \int_C \frac{\varphi_0'(z)}{\varphi_0(z) - \lambda} dz$$

makes sense and is the number of times the function φ_0 takes the value λ within the curve C, and is therefore at least 2. However, this integral is evidently the limit of the numbers

$$W(\lambda, \varphi_n, C) = \frac{1}{2\pi i} \int_C \frac{\varphi_n'(z)}{\varphi_n(z) - \lambda} dz$$

since the integrands converge uniformly on C for n sufficiently large, and these numbers can only be 0 or 1 since the functions φ_n are univalent. Thus φ_0 is also univalent.

Now suppose that λ is a point in the image of φ_0. There exists a point z_0 such that $\varphi_0(z_0) = \lambda$ and a small circle C in the domain of φ_0

with z_0 as center. From the convergence of $W(\lambda, \varphi_n, C)$ to $W(\lambda, \varphi_0, C)=1$ we infer that λ is in the image of φ_n for large enough n. It follows that Ω_0 is a subset of the union of the Ω_n.

It is a little harder to show that Ω_0 coincides with that union. Let K denote the closed disk $|z| \leqq \frac{1}{2}$ and D the open disk $|z|<1$. Let C be the circle $|z|=3/4$. Since the functions φ_n converge uniformly on C, the closed curves $\varphi_n(C)$ converge to $\varphi_0(C)$ and it follows that for large enough n $\varphi_0(K)$ is contained in $\varphi_n(D)$. Therefore the functions $F_n(z) = \varphi_n^{-1}(z)$ are bounded and analytic in a neighborhood of $\varphi_0(K)$, and we can extract a convergent subsequence. Suppose the subsequence converges to $F(z)$ on $\varphi_0(K)$. That set contains a neighborhood of 0, and in particular, an interval on the real axis of the form $(-b, b)$ for some $b>0$. An elementary consideration of the graphs of the functions $\varphi_n(x)$ and $\varphi_n^{-1}(x)$ on that interval makes it obvious that the sequence $F_n(x)$ converges on that interval to $\varphi_0^{-1}(x)$. This identifies the function $F(z)$, which must be an analytic continuation of $\varphi_0^{-1}(z)$ and is therefore uniquely determined. Hence the original sequence $\varphi_n^{-1}(z)$ converges to $\varphi_0^{-1}(z)$ on $\varphi_0(K)$, and indeed, on a neighborhood of the closed interval $[-1, 1]$.

We now fix an integer N and for $n>N$ we consider the sequence of Pick functions $\varphi_n^{-1} \circ \varphi_N$. On a neighborhood of 0 this sequence converges to $\varphi_0^{-1} \circ \varphi_N$ and so in particular it is bounded at some point in the upper half-plane. From Lemma 3 of Chapter II we infer the existence of a subsequence converging in P, and therefore of a subsequence of φ_n^{-1} converging on Ω_N to some function $F(z)$. We have already identified $F(z)$ as an analytic continuation of $\varphi_0^{-1}(z)$ and so Ω_N is contained in the image of φ_0 since the set is simply connected and the analytic continuation of $\varphi_0^{-1}(z)$ is unambiguous. Since the N is arbitrary, this proves the lemma.

Lemma 2. *Let φ_n be a sequence in $\mathcal{S}$ such that the images Ω_n form a decreasing sequence of sets. Then the sequence converges in P to a univalent function φ_0 with image Ω_0 where Ω_0 is the component of the interior of $\bigcap_n \Omega_n$ containing the origin.*

Since the proof differs only trivially from that of Lemma 1 we omit it. Note that we do not know that φ_0 is in $\mathcal{S}$ since there may not be a value $a>1$ so that φ_0 belongs to $P(-a, a)$.

Consider an arbitrary function φ of the class $\mathcal{S}$ with image Ω. It is obvious that there exists an increasing sequence of bounded open sets Ω_n with union Ω such that every Ω_n is simply connected. We may also require that the boundary of Ω_n be a polygonal curve which intersects the real axis in a closed interval $[x_-, x_+]$ with $x_- < -1$ and $x_+ > 1$.

There will correspond a sequence φ_n in $\mathscr{S}$ with Ω_n the image of φ_n, and in view of Lemma 1 this sequence converges in P to φ. It follows that the set of functions in $\mathscr{S}$ having images bounded by such polygonal curves is dense in $\mathscr{S}$. Suppose next that φ_1 is such a function with image Ω_1. The boundary of Ω_1 is a polygonal curve which intersects the real axis in the interval $[x_-, x_+]$. We parameterise the part of the boundary in the upper half-plane by a continuous function $\gamma(t)$ on $0 \leqq t \leqq 1$ putting $\gamma(0) = x_+$ and $\gamma(1) = x_-$. Now let Γ_t be the image of the interval $0 \leqq s \leqq t$ under γ and let Ω_t be the complement of Γ_t in the upper half-plane. There exists a uniquely determined function φ_t in $\mathscr{S}$ having Ω_t as its image. The sets Ω_t form a decreasing family and from Lemma 2 it follows that the corresponding φ_t converge in P to φ_1. From this it is clear that the set $\mathscr{S}_0$ of functions in $\mathscr{S}$ having images which are complements of Jordan arcs in the half-plane issuing from some real point $x_+ > 1$ is a dense subset of that class. In the balance of this chapter we will be concerned only with functions in $\mathscr{S}_0$ and so we now turn to the canonical integral representation of such functions.

Let φ be in $\mathscr{S}_0$ with image Ω and let x_+ be that end point of the corresponding Jordan arc Γ which lies on the real axis. The function maps the upper half-plane conformally on Ω. From a theorem of Carathéodory in the theory of conformal mapping it follows that φ may be extended to a homeomorphism of the closed upper half-plane onto the closure of Ω provided certain conventions are made. We must regard every point of Γ other than the non-real end point as two points, depending on which side of Γ the point is approached from. We must also make the usual conventions concerning the point at infinity. We can now easily identify the measure μ occuring in the canonical representation of φ: as y approaches 0 the Pick function

$$\varphi(x+iy) = U(x+iy) + iV(x+iy)$$

converges to a finite value for every x, except that x which goes into the point at infinity. Such an x is a pole of the function and corresponds to a point mass in the measure. It may happen, of course, that the pole is at infinity, and this will be the case if and only if the coefficient α is positive. Elswhere the measure will be of the form

$$d\mu(\lambda) = \frac{1}{\pi} V(\lambda) d\lambda$$

where $V(\lambda)$ is continuous and bounded on the real axis. The set where $V(\lambda)$ is not zero is the open interval of the axis which the homeomorphism φ carries into the non-real points of Γ. This interval, of course, may contain the point at infinity, or have that point as an end point.

The end points of the interval are the two points on the axis which φ carries into x_+. We call x_+ the exceptional real value of the function and note that it is the only real point where φ^{-1} is not analytic.

It is our purpose in this chapter to show that functions φ in the class $\mathscr{S}_0$ can be obtained from a certain differential equation. We shall therefore suppose in the sequel that a function φ in $\mathscr{S}_0$ with image Ω corresponding to a Jordan arc Γ has been given once and for all, where the exceptional real value of the function is some $x_+ > 1$. To study φ in greater detail we introduce a special parameterization of the arc Γ. Let $\gamma(t)$ be defined on the interval $0 \leqq t \leqq T$ where $\gamma(0)$ is the endpoint of Γ lying in the upper half-plane, and where $\gamma(T)$ is the exceptional real value x_+ of Γ. By Γ_t we denote the image under γ of the closed interval $[t, T]$ and by Ω_t we mean the complement of Γ_t in the half-plane. As t approaches T the Ω_t increase to Ω_T, the whole upper half-plane. Let g_t be the family of corresponding functions in $\mathscr{S}_0$. Evidently these depend continuously on t and $g_0(z) = \varphi(z)$ while $g_T(z) = z$ for all z. It is important to notice that if $t < s$ then $\Omega_t \subset \Omega_s$ and therefore that the composition $H_{t,s} = g_s^{-1} \circ g_t$ exists and belongs to $\mathscr{S}$. Its derivative at the origin is

$$H'_{t,s}(0) = g'_t(0)/g'_s(0)$$

a number in the interval $(0, 1)$ since $H_{t,s}$ is in $\mathscr{S}$ and is not the identity mapping. Thus $g'_t(0) < g'_s(0)$ and the function $g'_t(0)$ is strictly monotone increasing in t. We are therefore permitted to make another convention and to suppose the parameterization so chosen that $g'_t(0) = e^{t-T}$. Somewhat later we will pass to another system of functions, namely the functions $\varphi_t = g_t^{-1} \circ g_0$, so that φ_0 will be the identity and φ_T the given function in $\mathscr{S}_0$.

Now we introduce another function. We consider each g_t as extended according to Carathéodory's theorem to the closed upper half-plane and let $\lambda(t)$ be the unique point which satisfies the equation $g_t(\lambda(t)) = \gamma(t)$. Let $I(t, s)$ be the smallest closed interval which supports the measure occuring in the canonical representation of

$$H_{t,s}(z) = g_s^{-1}(g_t(z))$$

where $t < s$. We should note that $\lambda(t)$ is an interior point of $I(t, s)$ and that $\lambda(s)$ is the exceptional real value of that function. The kernel of our argument is contained in the following lemma.

Lemma 3. *The function* $k(t) = 1/\lambda(t)$ *is continuous on* $0 \leqq t \leqq T$, *and as the points* t *and* s *approach a common limit* r *the intervals* $I(t, s)$ *converge to the point* $\lambda(r)$.

Proof. We first consider the harmonic measure of the interval $I(t,s)$: this is a function $V_{t,s}(z)$ harmonic in the upper half-plane taking the boundary values 0 on the complement of $I(t,s)$ and having the boundary value 1 on the interior of the interval. From the representation theorem of Chapter II we know that $V_{t,s}$ may be put in the form of an integral:

$$V_{t,s}(x+iy)=\frac{1}{\pi}\int\limits_{I(t,s)}\frac{y}{(\lambda-x)^2+y^2}\,d\lambda$$

and this integral can be computed explicitly to show that the value of $V_{t,s}$ at the point z in the half-plane is the angle under which $I(t,s)$ is seen from z divided by π. The function admits an extension to the closed half-plane which is continuous everywhere except at the endpoints of $I(t,s)$. These are the points where the extended $H_{t,s}(z)$ takes the value $\lambda(s)$. Now we pass to the set Ω_t and consider there the harmonic function

$$v_{t,s}=V_{t,s}\circ g_t^{-1};$$

since g_t^{-1} can be extended to be a homeomorphism of the closure of Ω_t onto the closed half-plane, we see that $v_{t,s}$ is continuous on the closure of Ω_t except at points which g_t^{-1} carries into the discontinuities of $V_{t,s}$. We can identify those points as follows: if $H_{t,s}(a)=\lambda(s)$, then $g_t(a)=g_s(\lambda(s))=\gamma(s)$ and $a=g_t^{-1}(\gamma(s))$. Thus $V_{t,s}$ is discontinuous at a and $v_{t,s}$ is discontinuous at $\gamma(s)$. Now $v_{t,s}(z)$ is continuous throughout the closed upper half-plane except at the point $\gamma(s)$; the function vanishes on that part of the Γ_t which is the image of the open interval (s,T) under γ and equals 1 on the rest of Γ_t, except, of course, for the point $\gamma(s)$.

Suppose t and s converge to a common limit r. Let G be a linear fractional transformation mapping the upper half-plane onto the disk $|z|<1$ which carries the point $\gamma(r)$ into the origin. Now for some small positive ε put

$$U_\varepsilon(z)=\frac{\log|z|}{\log\varepsilon}$$

and pass to the composition $W_\varepsilon=U_\varepsilon\circ G$. Let C_ε be the circle surrounding $\gamma(r)$ which is the image of $|z|=\varepsilon$ under G^{-1}. The function $W_\varepsilon(z)$ is harmonic in the half-plane except at $\gamma(r)$ and equals 1 on the circle C_ε. Moreover, this function admits a continuous extension to the real axis where it vanishes.

We suppose now that t and s are so close to r that the arc between $\gamma(t)$ and $\gamma(s)$ is wholly surrounded by the circle C_ε. Let D_ε be the disk with boundary C_ε and $\Omega^*=\Omega_t-D_\varepsilon$. The function

$$W_\varepsilon(z)-v_{t,s}(z)$$

is harmonic in Ω^* and admits a continuous extension to the boundary, upon which it is nonnegative. It follows that $v_{t,s}(z) \leqq W(z)$ for all z in Ω^* and in particular for all z in Ω_0. As ε approaches 0 the function $U_\varepsilon(z)$ evidently converges to 0 uniformly on compact subsets of the disk

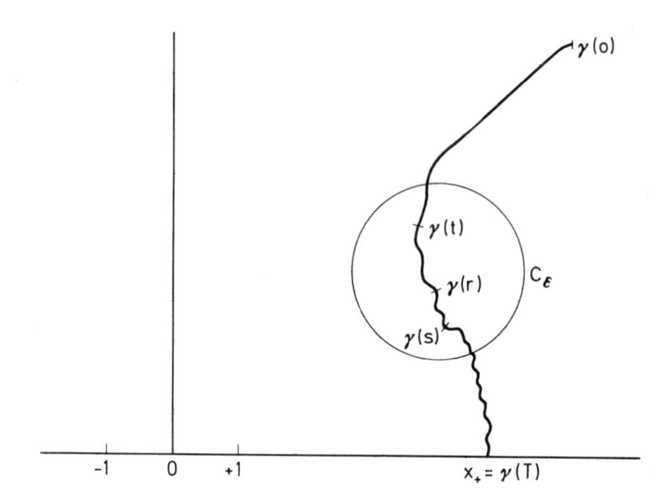

$|z|<1$ which do not contain the origin. Thus $W_\varepsilon(z)$ converges to 0 uniformly on compact subsets of Ω_0 and so $v_{t,s}(z)$ also does. Let z_0 be chosen in such a way that the compact set K determined by $g_t(z_0)$ as t runs through the interval $[0, T]$ does not intersect the arc Γ; as t and s approach r the functions $v_{t,s}(z)$ converge to 0 uniformly on K and so $V_{t,s}(z_0)$ converges to 0. If the functions $V_{t,s}$ converge to 0 at one point, they evidently do so for all points in the half-plane. We infer finally that the angle under which $I(t,s)$ is seen from any fixed point z in the upper half-plane converges to 0 as t and s approach a common limit.

We next observe that as t and s approach one another the functions $H_{t,s}(z)$ converge in $P(-1,1)$ to the identity mapping, since

$$H'_{t,s}(0) = e^{t-s}$$

converges to 1 and all of the functions are in the class $\mathscr{S}$. The measure corresponding to $H_{t,s}$ consists of a point mass, perhaps at infinity, and an absolutely continuous measure supported by $I(t,s)$. It is easy to verify that as t and s approach a common limit the possible point mass in the measure moves out to infinity, and if the interval (a,b) is disjoint from the system of intervals $I(t,s)$, then $H_{t,s}(z)$ converges in $P(a,b)$ to $\varphi_0(z) = z$.

Our argument to show the continuity of $k(t) = 1/\lambda(t)$ depends on the convergence of $H_{t,s}$ to the identity on intervals (a,b) disjoint from the support of the measure, and also on another circumstance. We may extend the definition of $H_{t,s}(x)$ to $I(t,s)$ by putting it equal to $\lambda(s)$ on that interval: the extended function is now continuous and monotone on the real axis provided $\lambda(s)$ is finite.

Suppose first that t is held fixed and that $\lambda(t)$ is finite. Let s diminish to t and consider the extended functions $H_{t,s}(x)$ on some interval (a,b) containing $\lambda(t)$. For s sufficiently close to t the interval $I(t,s)$, which contains $\lambda(t)$, is wholly contained in (a,b). Now

$$H_{t,s}(a) < H_{t,s}(\lambda(t)) = \lambda(s) < H_{t,s}(b)$$

and $H_{t,s}(a)$ converges to a while $H_{t,s}(b)$ converges to b. Since (a,b) was an arbitrary interval containing $\lambda(t)$ we infer that $\lambda(s)$ converges to $\lambda(t)$. This means that $\lambda(t)$ is continuous on the right at any point where it is finite, or that $k(t)$ is continuous on the right where it is not zero.

Now suppose that $\lambda(t)$ is the point at infinity, i.e., $k(t)=0$, and that as s approaches t through some sequence of values the numbers $\lambda(s)$ approach a finite limit L. Consider the interval (a,b) where $a=L-2\varepsilon$ and $b=L+2\varepsilon$ for some positive ε. The numbers $H_{t,s}(a)$ converge to a while $H_{t,s}(b)$ converges to b, so for s sufficiently close to t the function $H_{t,s}(x)$ takes on all values in $(L-\varepsilon, L+\varepsilon)$ inside (a,b). This clearly implies that (a,b) has a non-empty intersection with $I(t,s)$, an interval which shrinks to the point at infinity, a contradiction. Hence $\lambda(s)$ can approach no finite limit, and so $k(t)$ is continuous on the right even when it vanishes.

We next hold s fixed and let t increase to s. The interval $I(t,s)$ is the image under g_s^{-1} of that part of the arc Γ between $\gamma(s)$ and $\gamma(t)$, and as t approaches s these arcs diminish monotonically. It follows that $I(t,s)$ shrinks to $g_s^{-1}(\gamma(s))=\lambda(s)$. Because $\lambda(t)$ is contained in $I(t,s)$ we see that $\lambda(t)$ converges to $\lambda(s)$, even when $\lambda(s)$ is the point at infinity. This means that $k(t)$ is continuous on the right, hence continuous on the interval $0 \leqq t \leqq T$.

Finally, if t and s converge to a common limit r, the intervals $I(t,s)$ shrink to a point, possibly the point at infinity, and these intervals all contain $\lambda(t)$. Thus $I(t,s)$ shrinks to $\lim \lambda(t)=\lambda(r)$, as asserted in the lemma. This completes the proof.

We will now be able to complete the argument of this chapter after a short calculation. Let $\varphi_0(z)$ be the identity mapping and consider the function $H_{t,s}(z)-\varphi_0(z)$. This function is analytic in a neighborhood of the closed interval $[-1,1]$ and vanishes at $-1, 0$ and 1. We use these facts to obtain a convenient integral representation for $H_{t,s}(z)-\varphi_0(z)$. Let α, β and μ be the elements occuring in the canonical integral representation of the Pick function $H_{t,s}(z)$; we will then have

$$H_{t,s}(z) - \varphi_0(z) = \alpha z - z + \beta + \int \left[\frac{1}{\lambda - z} - \frac{\lambda}{\lambda^2 + 1} \right] d\mu(\lambda)$$

and since the function vanishes at the origin we infer that

$$\beta = \int \left[\frac{-1}{\lambda} + \frac{\lambda}{\lambda^2 + 1} \right] d\mu(\lambda) .$$

Now

$$H_{t,s}(z) - \varphi_0(z) = (\alpha - 1) z + \int \left[\frac{1}{\lambda - z} - \frac{1}{\lambda} \right] d\mu(\lambda)$$

and because this vanishes at $z = 1$

$$0 = (\alpha - 1) + \int \left[\frac{1}{\lambda - 1} - \frac{1}{\lambda} \right] d\mu(\lambda)$$

whence

$$\alpha - 1 = - \int \frac{d\mu(\lambda)}{\lambda(\lambda - 1)}$$

and

$$\begin{aligned} H_{t,s}(z) - \varphi_0(z) &= \int \left[\frac{-z}{\lambda(\lambda - 1)} + \frac{1}{\lambda - z} - \frac{1}{\lambda} \right] d\mu(\lambda) \\ &= \int \frac{z^2 - z}{\lambda(\lambda - 1)(\lambda - z)} d\mu(\lambda) . \end{aligned}$$

For $z = -1$ we have

$$0 = 2 \int \frac{d\mu(\lambda)}{\lambda(\lambda - 1)(\lambda + 1)}$$

and so the function

$$N(z) = \int \frac{z^2 - z}{\lambda(\lambda^2 - 1)} d\mu(\lambda)$$

vanishes identically.

It follows that

$$\begin{aligned} H_{t,s}(z) - \varphi_0(z) &= \int \left[\frac{z^2 - z}{\lambda(\lambda - 1)(\lambda - z)} - \frac{z^2 - z}{\lambda(\lambda^2 - 1)} \right] d\mu(\lambda) \\ &= \int \frac{z^3 - z}{(\lambda - z)\lambda(\lambda^2 - 1)} d\mu(\lambda) . \end{aligned}$$

If we finally take the derivative at the origin we obtain

$$H'_{t,s}(0) - 1 = e^{t-s} - 1 = - \int \frac{d\mu(\lambda)}{\lambda^2(\lambda^2 - 1)} .$$

We introduce a new positive measure, defining

$$d\nu_{t,s}(\lambda) = \frac{d\mu(\lambda)}{(s-t)\lambda^2(\lambda^2-1)}$$

noting that

$$\int d\nu_{t,s}(\lambda) = \frac{1-e^{t-s}}{s-t} = 1-(s-t)+\frac{(s-t)^2}{6}\cdots.$$

Thus the total mass of the measure converges to 1 as t and s approach a common limit. We may finally write

$$\frac{H_{t,s}(z)-\varphi_0(z)}{s-t} = (z^3-z)\int \frac{\lambda}{\lambda-z}\,d\nu_{t,s}(\lambda).$$

The measure $\nu_{t,s}$ is supported by the interval $I(t,s)$ and the possible pole of $H_{t,s}(z)$. As t and s approach a common limit the pole moves off to infinity and the interval $I(t,s)$, as we have seen, shrinks to a point, which may be the point at infinity. Let r be the common limit of t and s: if $\lambda(r)$ is the point at infinity the quotient

$$\frac{H_{t,s}(z)-\varphi_0(z)}{s-t}$$

converges to z^3-z. In general, since the system of quotients is a family of analytic functions, bounded on compact subsets of the plane, there will at least exist suitable convergent subsequences, and for such a subsequence we will obtain a limit of the form

$$(z^3-z)\left[\frac{p(r)}{1-k(r)z}+q(r)\right]$$

where $k(r)=1/\lambda(r)$ is continuous and $p(r)$ and $q(r)$ are in the interval $[0,1]$ with $p(r)+q(r)=1$.

Next we turn to the functions $\varphi_t = g_t^{-1}\circ g_0$ already introduced and observe that

$$H_{t,s}\circ\varphi_t = g_s^{-1}\circ g_t\circ g_t^{-1}\circ g_0 = g_s^{-1}\circ g_0 = \varphi_s,$$

hence

$$\frac{\varphi_s-\varphi_t}{s-t} = \frac{H_{t,s}\circ\varphi_t-\varphi_t}{s-t} = \frac{H_{t,s}-\varphi_0}{s-t}\circ\varphi_t.$$

We have already observed that the difference quotients $(H_{t,s}-\varphi_0)/(s-t)$ are uniformly bounded on compact subsets of the upper half-plane; it will then follow that for any fixed z in the upper half-plane the quotients $(\varphi_s(z)-\varphi_t(z))/(s-t)$ are uniformly bounded, and therefore that $\varphi_t(z)$ is Lipschitzian in the real parameter t. This means that the function is differentiable almost everywhere and we have finally, almost everywhere in t,

$$\frac{d\varphi_t(z)}{dt} = (\varphi_t(z)^3 - \varphi_t(z))\left[\frac{p(t)}{1-k(t)\varphi_t(z)} + q(t)\right].$$

It is easy to show that the coefficient $p(t)$, and therefore $q(t)$ also, is a measurable function of t. If $P(t,s)$ is the $\nu_{t,s}$-measure of the interval $I(t,s)$, then $P(t,s)$ is a continuous function in the triangle $s>t$ and $p(t)$ is almost everywhere a limit of $P(t,s)$ as s approaches t. Not only does this show that $p(t)$ is measurable, it also shows, since φ_t is Lipschitzian, hence absolutely continuous, that for almost all t only one limit is possible i.e., the functions $(H_{t,s}(z)-\varphi_0(z))/(s-t)$ actually converge to a limit for almost all t. Thus the differential equation satisfied by φ_t is independent of the initial value z at $t=0$. These considerations are also valid for z in the lower half-plane or on the interval $[-1,1]$.

Chapter XIX. More Analytic Continuation

In this chapter we first study solutions to the differential equation

$$\frac{dy}{dt} = \eta(y, t)$$

where the continuous function $\eta(y, t)$ is defined for t in the closed interval $[0, T]$ and y in the open interval (c, d). We consider the differential equation relative to an interval (a, b) where $c<a<b<d$ and with the initial condition $y(0, x)=x$. If we suppose that $\eta(y, t)$ is continuously differentiable in y the standard existence and uniqueness theorems guarantee that $y(t, x)$ as a function of x is strictly monotone and continuously differentiable. We write

$$y(t, x) = f_t(x)$$

and note that this function is defined on (a, b) for all sufficiently small t.

The principal hypothesis we make concerning the function $\eta(y, t)$ is that for a given integer $n>1$ and for any choice of n distinct points λ_i in (c, d) and any fixed value of t the corresponding Pick matrix of order n

$$[\lambda_i, \lambda_j]_\eta = \frac{\eta(\lambda_i, t) - \eta(\lambda_j, t)}{\lambda_i - \lambda_j} \quad \text{for } i \neq j,$$

$$[\lambda_i, \lambda_i]_\eta = \eta'(\lambda_i, t)$$

is an almost positive matrix. Accordingly, $\eta(y, t)$ belongs to $Q_n(c, d)$ for all t.

Theorem I. *If the solution $f_t(x)$ is defined on (a, b) for all t in an interval $[0, T_0]$ then those functions belong to the class $P_n(a, b)$.*

Proof. The function $f_t(x)$ belongs to $P_n(a, b)$ if and only if for every choice of n distinct points λ_i in the interval the corresponding Pick matrix is a positive matrix. We write that matrix $K(t) = k_{ij}(t)$ where

$$k_{ij}(t) = \frac{y(t, \lambda_i) - y(t, \lambda_j)}{\lambda_i - \lambda_j} \quad \text{for } i \neq j$$

and

$$k_{ii}(t)=y'(t,\lambda_i).$$

For $i\neq j$ we can differentiate with respect to t to obtain

$$\frac{dk_{ij}(t)}{dt}=\frac{\eta(y(t,\lambda_i),t)-\eta(y(t,\lambda_j),t)}{\lambda_i-\lambda_j}$$

and since $y(t,x)$ is strictly increasing in x, this may be written

$$\frac{\eta(y(t,\lambda_i),t)-\eta(y(t,\lambda_j),y)}{y(t,\lambda_i)-y(t,\lambda_j)}\cdot\frac{y(t,\lambda_i)-y(t,\lambda_j)}{\lambda_i-\lambda_j}$$

whence

$$\frac{dk_{ij}(t)}{dt}=\frac{\eta(y(t,\lambda_i),t)-\eta(y(t,\lambda_j),t)}{y(t,\lambda_i)-y(t,\lambda_j)}k_{ij}(t).$$

We can argue similarly with the diagonal elements to find that

$$\frac{dk_{ii}(t)}{dt}=\frac{\partial\eta}{\partial y}(y(t,\lambda_i),t)k_{ii}(t).$$

For $t=0$ we have $k_{ij}(0)=1$, even when $i=j$ and it is therefore easy to find the solution to the differential equations obtained above. Let $A(t)=a_{ij}(t)$ be the Pick matrix of the function $\eta(y,t)$ associated with the system of n distinct points $y(t,\lambda_i)$. We will then have

$$k_{ij}(t)=\exp\int_0^t a_{ij}(s)\,ds$$

even for $i=j$. By hypothesis the matrix $A(s)$ is almost positive for all s in the interval $[0,t]$ and so the integral above, as the limit of linear combinations with positive coefficients of almost positive matrices, is again almost positive. The Schur exponential is therefore a positive matrix. This puts $y(t,x)=f_t(x)$ in the class $P_n(a,b)$, as desired.

Let us note that the condition of the theorem is a necessary one if the function $\eta(y,t)$ is independent of t. For if the solutions $f_t(x)$ with $f_0(x)=x$ are all in $P_n(a,b)$ for sufficiently small values of t, then their corresponding Pick matrices $K(t)$ computed for some fixed but arbitrary set of n distinct points λ_i in the interval are positive matrices. Since every entry in $K(0)$ is $+1$, the matrix $-K(0)$ is almost positive and therefore $K(t)-K(0)$ is almost positive as well. It follows that the matrices

$$\frac{K(t)-K(0)}{t}$$

are also almost positive, and these converge with diminishing t to the Pick matrix of $\eta(x)$ computed for the points λ_i.

It is an evident consequence of Theorem I that if the function $\eta(y,t)$ has an almost positive Pick matrix for all choices of the n distinct points λ_i and all values of n then the solutions $y(t,x)=f_t(x)$ are in fact in $P(a,b)$. In this case, both the solutions and the functions $\eta(y,t)$ admit an analytic continuation in the half-plane for fixed t. We use this fact in the proof of the following theorem of FitzGerald.

Theorem II. *Let $f(x)$ be a real, strictly increasing C^2-function on an interval (a,b) containing the origin such that $f(0)=0$. Let $\eta(x)$ be the C^1-function $-f(x)/f'(x)$. Then $f(x)$ is a univalent function in $P(a,b)$ with a range star-shaped relative to the origin if and only if the Pick matrix of $\eta(x)$ is almost positive, i.e. if and only if $\eta(x)$ belongs to $Q(a,b)$.*

Proof. First suppose that $f(x)$ is an element of $P(a,b)$ that is univalent and has a range star-shaped relative to 0. We form a semi-group of functions in $P(a,b)$ by defining

$$\varphi_t(z)=f^{-1}(e^{-t}f(z)).$$

The definition makes sense since the range of the function is star-shaped relative to 0. It is easy to see that $\varphi_t(a)$ is in $P(a,b)$ and that these functions form a semi-group under composition:

$$\varphi_{t+s}=\varphi_t\circ\varphi_s, \qquad \varphi_0(z)=z.$$

The semi-group depends differentiably on t, and we have

$$\frac{d\varphi_t}{dt}=\eta\circ\varphi_t.$$

The functions $\varphi_t(x)$ being in $P(a,b)$, they belong to all the classes $P_n(a,b)$ and the remarks following theorem I make it clear that $\eta(x)$ has an almost positive Pick matrix.

If, on the other hand, we suppose that $\eta(x)$ has an almost positive Pick matrix, then the results of Chapter XV guarantee that the function has an analytic continuation throughout the domain $\mathscr{D}$ consisting of the union of the two open half-planes and the interval (a,b). From Theorem III of that chapter we know that the function is of the form

$$\eta(x)=Ax+B+x^2\varphi(x)$$

where $\varphi(x)$ is in $P(a,b)$. However, it is easy to verify that $\eta(0)=0$ and that $\eta'(0)=-1$, whence

$$\eta(x)=-x+x^2\varphi(x).$$

For all z in $\mathscr{D}$, then

$$\eta(z)=z^2[\varphi(z)-1/z]$$

where the function in brackets is a non-trivial Pick function. We infer that $\eta(z)$ has only one zero in $\mathscr{D}$ and that a simple zero at the origin. It follows that the negative reciprocal of $\eta(z)$ is analytic in $\mathscr{D}$ except for a simple pole at the origin. From the behavior of $\eta(z)$ near 0 we can verify that the residue at that pole is $+1$, so that

$$\frac{-1}{\eta(z)} = h(z) + 1/z$$

where $h(z)$ is analytic in $\mathscr{D}$. We now form the function

$$L(z) = \int_c^z h(\zeta)\, d\zeta + \log z + \log(f(c)/c)$$

where $c = b/2$. Evidently this is of the form $H(z) + \log z$ where $H(z)$ is analytic in $\mathscr{D}$, and taking exponentials we find

$$F(z) = \exp L(z)$$

analytic in $\mathscr{D}$ and real on (a, b). The logarithmic derivatives of $F(x)$ and $f(x)$ coincide on the interval (a, b) and therefore, on $(0, b)$, $F(x) = C f(x)$ for some constant C. However, the equation $F(c) = f(c)$ shows that $C = 1$. Similarly, for the interval $(a, 0)$ we have $F(x) = C f(x)$, and since both functions are differentiable at 0 this implies $F'(0) = C f'(0)$, and we again infer that $C = 1$ since $F(x) = f(x)$ to the right of the origin. In this way we have obtained an analytic continuation of $f(x)$ to $F(z)$ defined in $\mathscr{D}$.

Now on the interval (a, b) consider the functions

$$\varphi_t(x) = f^{-1}(e^{-t} f(x))$$

which are well defined since $f(x)$ has an inverse as a real function. These functions form a semi-group:

$$\varphi_{t+s}(x) = \varphi_t(\varphi_s(x)), \qquad \varphi_0(x) = x\,.$$

It is also clear that with increasing t the functions $\varphi_t(x)$ converge to 0 uniformly on compact subsets of the interval (a, b). The semi-group depends differentiably on t, and the differential equation

$$\frac{d\varphi_t}{dt} = \eta(\varphi_t)$$

is satisfied. Since the function $\eta(x)$ has an almost positive Pick matrix it follows that the functions $\varphi_t(x)$ are in $P(a, b)$ and by analytic continuation the semi-group property of the family holds throughout $\mathscr{D}$:

$$\varphi_{t+s}(z) = \varphi_t(\varphi_s(z)), \qquad \varphi_0(z) = z\,.$$

Moreover, as t increases to infinity the numbers $\varphi_t(z)$ converge to 0 for every z in $\mathscr{D}$.

Since $F(x)=f(x)$ on (a,b) we may write

$$F(\varphi_t(x))=e^{-t}f(x)$$

to infer by analytic continuation that

$$F(\varphi_t(z))=e^{-t}F(z)$$

for all points z in $\mathscr{D}$. Thus, for all positive t, $e^{-t}F(z)$ is in the range of F, and this shows that the range of that function is star-shaped relative to 0.

Since $F(z)$ has a positive derivative at the origin, there exists a small neighborhood N of 0 such that $F(z)$ is a homeomorphism of that neighborhood onto $F(N)$, another neighborhood of 0. The homeomorphism maps real points in N onto real points in $F(N)$ and carries points of N with a positive imaginary part into images with the same property.

To show that $F(z)$ is in $P(a,b)$ we have only to show that points in the upper half-plane are mapped into points in the same half-plane. If z is a point in the upper half-plane such that $F(z)$ is in the lower half-plane, then for sufficiently large t the point $\varphi_t(z)$ is in the neighborhood N and has a positive imaginary part, since φ_t is in $P(a,b)$. The image of this point under F is $e^{-t}F(z)$ and is therefore in the lower half-plane. This contradicts the known behavior of $F(z)$ on the set N.

We argue similarly to show that $F(z)$ is univalent: if there exist two points z_1 and z_2 in the upper half-plane having the same image under F, then for large enough t the points $\varphi_t(z_1)$ and $\varphi_t(z_2)$ are in N and have the same image in $F(N)$. Because F is a homeomorphism on N this means that the numbers $\varphi_t(z_1)$ and $\varphi_t(z_2)$ are equal. We complete the proof by showing that this does not happen.

The number $\varphi_t(z_1)$ depends continuously on t, and as t increases runs through a trajectory issuing from $\varphi_0(z_1)=z_1$. Let s be the smallest value of t for which $\varphi_t(z_1)=\varphi_t(z_2)$. Evidently $s>0$ since z_1 and z_2 are distinct. Let z_0 be $\varphi_s(z_1)$, the point where the two trajectories intersect, and let D be a circular neighborhood of z_0 lying wholly in the upper half-plane. For small values of h the points $\varphi_{s-h}(z_1)$ and $\varphi_{s-h}(z_2)$ are distinct points in D which the map φ_h carries into z_0. For such small positive values of h, then, the integral

$$\frac{1}{2\pi i}\int_C \frac{\varphi_h'(z)}{\varphi_h(z)-z_0}\,dz$$

where C is the boundary of D is a positive integer at least as large as 2 since the integral equals the number of times the function $\varphi_h(z)$ takes the value z_0 in D. As h tends to 0 the integrand approaches $1/(z-z_0)$ uniformly on C and the integral converges to $+1$. This contradiction completes the proof.

Notes and Comment

Chapter I. The standard work on divided differences is Nörlund 1. The theorem of Schur first appeared in Schur 1. The criterion for positivity occurs in Carathéodory 4. In a later chapter we will need a more general form of this criterion in which det A may equal 0. An analysis of our argument shows that it goes through in this case also: if $\lambda_1=0$ and λ_2 is negative, then again there is a subspace of dimension 2 on which the quadratic form (Au, u) is non-positive, and that space contains a normalized vector u with $u_{n+1}=0$. It follows as before that (Au, u) is positive, a contradiction. Regularization is a standard tool of analysis; a detailed discussion can be found, for example, in Donoghue 2. The completely monotone functions were studied by S. N. Bernstein in Bernstein 1 but our proof is taken from Feller 1.

Chapter II. The Pick functions are discussed at some length in Stone 1 and also in Aronszajn and Donoghue 1 and 2. The integral representation seems first to have been established as our Theorem III by Herglotz in Herglotz 1. Much earlier, however, Stieltjes had introduced the Stieltjes integral in order to have integral representations for functions closely related to Pick functions. See Stieltjes 1. Physicists often call the negative of a function in $P(0, +\infty)$ a "series of Stieltjes".

Chapter III. Theorem I is part of a theorem of Pick to be found in Pick 2. We have had an instructive conversation with G. Choquet concerning the theorem of Hindmarsh, given in a more general form in Hindmarsh 1. The remaining theorems of the chapter are essentially due to Loewner and can be found implicit in Löwner 1 although some of our proofs appear in Donoghue 4. The extended Loewner matrix seems first to have been used by a student of Loewner in Dobsch 1.

Chapter IV. The Fatou theorem was first established in Fatou 1 and is extensively studied in Loomis 1 where most of the theorems of this chapter occur. Our proof of Theorem II is essentially that given by L. Hörmander in a private conversation. The proof of Theorem III appears in Donoghue 1 where an instructive conversation with A. Robertson was not acknowledged; the theorem in question is quite old. The Wiener Tauberian theorem can be found in Wiener 1 and also in Donoghue 2. The boundary behavior of a Pick function is studied in detail in Aronszajn and Donoghue 1 and 2 as well as Aronszajn 2.

Chapter V. The content of this chapter is classical; most of the results are in Stone 1 and in a somewhat more modern presentation in Dunford and Schwartz 1. The standard proof of the spectral theorem for bounded operators is given in Eberlein 1. See also Halmos 1 for a discussion including the case of non-separable spaces.

Chapter VI. The study of how the eigenvalues of a completely continuous self-adjoint operator change under a finite dimensional perturbation was doubtless initiated by H. Weyl in Weyl 2 some of whose theorems are reported in Riesz-Nagy 1. Many of the devices of this chapter occur in Aronszajn 1. See also Gould 1 and Donoghue 3.

Chapter VII. Most of the results of this chapter are in Löwner 1 although Theorems III and IV are in Dobsch 1.

Chapter VIII. The trace class of operators is studied at length in Schatten 1. Most of the theorems of this chapter are in Löwner 1 as well as Loewner 1.

Chapter IX. Loewners Theorem first occurs in Löwner 1. The argument given here is largely taken from Bendat and Sherman 1. The Hamburger moment problem is studied in detail in Shohat and Tamarkin 1 and also in Widder 1. An important study of Loewners Theorem from the viewpoint of the physicist can be found in Wigner and von Neumann 1.

Chapter X. The kernel function was introduced by S. Bergman in 1922 while the theory of reproducing kernels is almost entirely the work of N. Aronszajn, presented in Aronszajn 3. An excellent book on the subject is Meschkowski 1 although that author does not follow our convention about sesqui-analyticity. A whole issue of the Rocky Mountain Journal of Mathematics in 1972 has been devoted to modern aspects of the theory and this issue contains an extremely readable account of the kernel function by E. Hille. See Hille 1 and also Bergman 1.

Chapter XI. The Pick Theorem appears in Pick 2. The Nagy-Koranyi proof of Loewners Theorem is in Koranyi 1 and Sz.-Nagy 1 and those authors have generalized and extended their results in Koranyi and Sz.-Nagy 1.

Chapter XII. The principal results of this chapter are in Löwner 1 but the content of the chapter must be of a high antiquity. Interpolation by rational functions was exhaustively studied by Cauchy and Jacobi who doubtless knew what we have given here. The standard book on rational approximation is Walsh 1. We should note the relation between the Cauchy interpolation problem and the problem of the Padé approximation of an analytic function $f(z)$. For every pair of positive integers (m, n) the corresponding element of the Padé table is a pair of polynomials $[\sigma(z); \tau(z)]$ with degree $\sigma(z) \leqq m$ and degree $\tau(z) \leqq n$ such that the ratio $\sigma(z)/\tau(z)$ has a McLaurin expansion coinciding with that of $f(z)$ in the first $m+n+1$ terms. Thus, in the case when there is only one interpolation point and $N=2n+1$ is odd, our theory is the theory of the diagonal elements of the Padé table. See Perron 1, Luke 1 and Wall 1. The study of Padé approximation is now very fashionable and a whole issue of the Rocky Mountain Journal of Mathematics in 1974 is given to the modern theory of Padé approximation.

Chapter XIII. It is appropriate here to refer to the Pick-Nevanlinna interpolation problem. We suppose that an infinite sequence z_n is given in the upper half-plane and that the values of a Pick function on that sequence are prescribed: $\varphi(z_n)=w_n$. Theorem I of Chapter XI gives a necessary and sufficient condition that a solution to this interpolation problem exists; we suppose that condition satisfied and enquire about the uniqueness of the solution. The set of all Pick functions satisfying the first N equations is evidently non-empty and convex, and so the values taken by functions in this set at a given point z form a convex subset of the upper half-plane. It is easy to determine the shape of this set: a complex value w is in the set if and only if the Pick matrix of order $N+1$ computed with the first N values of z_n as well as $w=\varphi(z)$ is a positive matrix. In view of the criterion for positivity given in Chapter I, this will be the case if and only if the determinant of that matrix is non-negative. Now that determinant, as a function of w, is a real valued function of the form

$$d(w)=A w \bar{w}+B w+\overline{B w}+D$$

where A and D are real; by completing the square we find that $d(w)=0$ describes a circle. The arguments of this chapter show that this circle may be obtained in another way: if $d(w)=0$ there is only one solution to the interpolation problem, and this is rational, of lower degree. We must therefore have

$$w=\frac{\sigma_0(z)+t\,\sigma_\infty(z)}{\tau_0(z)+t\,\tau_\infty(z)}$$

for some appropriate real t. As t runs through the real axis the corresponding w traces out a circle in the half-plane.

If we now let N tend to infinity we obtain, for any fixed z in the half-plane, a sequence of circles C_N with C_{N+1} inside C_N. The solution to the interpolation problem is unique if and only if, for every z, these circles diminish to a point. Otherwise, in general, the circles will shrink to a non-trivial limiting circle. The two cases which may occur are therefore called limit-point and limit-circle, and this dichotomy arises in a number of problems in analysis. It was first observed by H. Weyl in Weyl 3 when he studied the spectral function of a singular Sturm-Liouiville operator. See Coddington and Levinson 1 or Aronszajn 2. The study of the uniqueness of the solution to the Hamburger moment problem also leads to a limit-point or limit circle case, as in Riesz 1 and Shohat and Tamarkin 1. This interpolation is also considered in Pick 1 and Nevanlinna 1 and 2. See also Weyl 1.

Most of the results of this chapter are in Löwner 1. In the special case that $v_i=0$ for all i the theorem is proved in Donoghue 4 in an entirely different way, making use of the elementary theory of convex sets. Those arguments could doubtless be generalized to obtain the results of this chapter as well.

Chapter XIV. The theorems of this chapter are mostly in Löwner 1. We should note that if $f(x)$ is a function in $P(a, b)$, hence in all of the classes $P_n(a, b)$, a pair of interpolation functions can give considerable information about $f(x)$. Thus if $\varphi(x)$ and $\hat{\varphi}(x)$ interpolate $f(x)$ for sets S and $\hat{S}$ corresponding to polynomials $S(x)$ and $\hat{S}(x)$, then on any subinterval of (a, b) where $S(x)$ and $\hat{S}(x)$ have opposite signs the function $f(x)$ is in between the two interpolating functions. We should also note that if S is a subset of $\hat{S}$ the interpolating function $\varphi(x)$ for $f(x)$ is also an interpolating function for $\hat{\varphi}(x)$ and hence Theorem III gives information about the relation between these two interpolating functions as well. It is therefore easy to determine sequences of sets S_n so that the corresponding interpolation functions converge monotonically to $f(x)$ in a given subinterval of (a, b).

Chapter XV. The almost positive matrices and the related infinitely divisible positive matrices first occured in the study of certain functions of positive type and corresponding infinitely divisible probability distributions. See Horn 1. We have insisted on the term "almost positive" to emphasise the intimate relation with the theory of reproducing kernels. Lemma 1 appears in many references, never with a proof, and was first explained to the author by J. Fehrman, whose proof was more geometric than that given in the text. Lemma 2 and Theorem I appear in Loewner 2 and essentially in Schoenberg 1. Theorem II is in Schoenberg 2 where it is attributed to Menger 1. Theorem III is in Horn 2. It seems probable that Q_n is not a local property for $n \geqq 3$ but we have not found an example to show this.

Chapter XVI. The content of Lemma 1 is classical; see, for example, Bergman 1 or Meschkowski 1. The analytic continuation of kernel functions was first studied by Bremmerman in Bremmerman 1 who showed, among other things, that if $K(z, w)$ was a Bergman kernel in $D \times D$ where D was open, and if the function admitted a sesqui-analytic continuation to a larger product domain $G \times G$, then the continuation was also a positive matrix. This result was greatly improved by FitzGerald, at least in the case of only one complex variable, so that $K(z, w)$ was required to be a positive matrix only on sets of lower dimension, e.g. the $I \times I$ of Lemma 2. FitzGeralds result is even a little more general than what we have attributed to him. See FitzGerald 1.The proof in the text for Lemma 2 is essentially that communicated to the author by L. Hörmander who has obtained the most general results in this direction. He has also provided the following instructive counterexample. Let $f(z)$ be an analytic function defined in an open set G in the space of one or more complex variables, and let N be the subset of G defined by the equation $f(z)=0$. Form the function $K(z, w) = -f(z)\overline{f(w)}$. This is evidently a positive matrix on $N \times N$ and is sesqui-analytic in $G \times G$, but there exists no set E properly containing N so that $K(z, w)$ is a positive matrix on $E \times E$.

Chapter XVII. The Loewner-FitzGerald theorem is proved in the union of the papers Loewner 2, Loewner 3 and FitzGerald 1. The proof given here is substantially simpler. Our comments about the differential equation $dy/dt = \theta(y, t)$ are sketchy, however, these remarks are not needed for the proof of the theorem. All that is required is that the Pick matrix for $\theta(y, t)$ be almost positive when t is fixed. Theorem II is essentially contained in a letter from FitzGerald to the author. Theorem III is due to Horn. See Horn 2 where a more general result is established.

Chapter XVIII. Lemma 1 and Lemma 2 are both special cases of a theorem of Carathéodory given in Carathéodory 3. See also Golusin 1 and Pfluger 1. Lemma 3 and its proof are largely taken from Pfluger 1. In Carathéodory 2 is found the theorem that a conformal mapping of the disk onto a region bounded by a Jordan curve can be extended to a homeomorphism of the closure. See Golusin 1, Seidel 1 and Pfluger 1 for example. The arguments in this chapter parallel those of Löwner given in Löwner 2 who considered a class of univalent functions mapping the disk $|z|<1$ into itself and leaving $z=0$ fixed. He also required the functions to have a positive derivative at the origin. Using the results of Carathéodory he could show that the subclass of those functions mapping the disk onto the complement of a Jordan curve connecting a boundary point with some interior point was dense in the compact open topology. He then obtained a differential equation satisfied by a family of functions corresponding to our $\varphi_t(z)$. This result he applied to obtain a proof of the Bieberbach conjecture for $n=3$, i. e. that if $f(z)$ was a function in the given class, its third McLaurin coefficient a_3 satisfied the inequality $|a_3| \leqq 3$ with equality only for the Koebe function. Löwners argument depended on the fact that he could derive differential inequalities for the coefficients from the differential equation. This result was obtained in 1923; over thirty years had to pass before further progress could be made with the Bieberbach conjecture. In 1955 Garabedian and Schiffer obtained the corresponding result for $n=4$ and their argument made essential use of Löwners differential equation. See Garabedian and Schiffer 1.

We should also note that we tacitly invoke the Riemann mapping theorem when we assume the existence of mapping functions $g_t(z)$ carrying the half-plane onto Ω_t.

Chapter XIX. Theorem I can be found in Loewner 2 and Theorem II is in FitzGerald 2 with quite a different proof.

Bibliography

Aronszajn, N.

1. Approximation methods for eigenvalues of completely continuous symmetric operators. Proceedings Symposium on Spectral Theory, p. 179–202 Oklahoma: Stillwater 1951.
2. On a problem of Weyl in the theory of singular Sturm-Liouiville equations. Amer. J. Math. **79**, 567–610 (1957).
3. Theory of reproducing kernels. Trans. Amer. Math. Soc. **68**, 337–404 (1950).

Aronszajn, N., Donoghue, W.

1. On exponential representations of analytic functions in the upper half-plane with positive imaginary part. J. Analyse Math. **5**, 321–385 (1956).
2. A supplement to the paper on exponential representations. J. Analyse Math. **12**, 113–127 (1964).

Bendat, J., Sherman, S.

1. Monotone and convex operator functions. Trans. Amer. Math. Soc. **79**, 58–71 (1955).

Bergman, S.

1. The kernel function and conformal mapping. New York: 1950

Bernstein, S.

1. Sur les fonctions absolument monotones. Acta Math. **52**, 1–66 (1928).

Bremmerman, H.

1. Holomorphic continuation of kernel function and the Bergman metric in serval complex variables. Lectures on functions of a complex variable. Ann Arbor: 1955.

Carathéodory, C.

1. Conformal representation. Cambridge: 1932.
2. Über die gegenseitige Beziehung der Ränder bei der konformen Abbildung des Innern einer Jordanschen Kurve auf einen Kreis. Math. Ann. **73**, 305–320 (1913).
3. Untersuchungen über die konforme Abbildung von festen und veränderlichen Gebieten. Math. Ann. **72**, 107–144 (1912).
4. Variationsrechnung. Berlin: 1935.

Coddington, E., Levinson, N.

1. Theory of ordinary differential equations. New York: 1955.

Dobsch, O.

1. Matrixfunktionen beschränkter Schwankung. Math. Z. **43**, 353–388 (1937).

Donoghue, W.
1. A theorem of the Fatou type. Monatsh. Math. **67**, 225–228 (1963).
2. Distributions and Fourier transforms. New York: 1969.
3. On the perturbation of spectra. Comm. Pure Appl. Math. **18**, 559–579 (1965).
4. The theorems of Loewner and Pick. Israel J. Math. **4**, 153–170 (1966).

Dunford, N., Schwartz, J.
1. Linear operators. Part II. New York: 1963.

Eberlein, W.
1. A note on the spectral theorem. Bull. Amer. Math. Soc. **52**, 328–331 (1946).

Fatou, P.
1. Séries trigonométriques et séries de Taylor. Acta Math. **30**, 335–400 (1906).

Feller, W.
1. On Muntz theorem and completely monotone functions. Amer. Math. Monthly **75**, 342–349 (1968).

FitzGerald, C.
1. On analytic continuation to a schlicht function. Proc. Amer. Math. Soc. **18**, 788–792 (1967).
2. On analytic continuation to a starlike function. Arch. Rational Mech. Anal. **35**, 397–401 (1969).

Garabedian, P., Schiffer, M.
1. A proof of the Bieberbach conjecture for the fourth coefficient. Arch. Rational Mech. Anal. **4**, 427–464 (1955).

Golusin, G.
1. Geometric theory of functions of a complex variable. Providence: 1969.

Gould, S.
1. Variational methods for eigenvalue problems. Toronto: 1957.

Halmos, P.
1. Introduction to Hilbert space and the theorey of spectral multiplicity. New York: 1951.

Herglotz, G.
1. Über Potenzreihen mit positivem, reelem Teil in Einheitskreis. Leipz. Bericht **63**, 501–511 (1911).

Hille, E.
1. Introduction to the general theory of reproducing kernels. Rocky Mountain Journal of Mathematics **2**, 321–368 (1972).

Hindsmarsh, A.
1. Picks conditions and analyticity. Pacific J. Math. **27**, 527–531 (1968).

Horn, R.
1. On infinitely divisible matrices, kernels and functions. Z. Wahrscheinlichkeitstheorie **8**, 219–230 (1967).
2. Schlicht mappings and infinitely divisible kernels. Pacific J. Math. **38**, 423–429 (1971).

Koranyi, A.
1. On a theorem of Löwner and its connections with resolvents of self-adjoint transformations. Acta Sci. Math. **17**, 63–70 (1956).

Koranyi, A., Sz.-Nagy, B.
1. Operatortheoretische Behandlung und Verallgemeinerung eines Problemkreises in der komplexen Funktionentheorie. Acta Math. **100**, 171–202 (1958).

Kraus, F.
1. Über konvexe Matrixfunktionen. Math. Z. **41**, 18–42 (1936).

Loewner, C.
1. On generation of monotonic transformations of higher order by infinitesimal transformations. J. Analyse Math. **11**, 189–206 (1963).
2. On Schlicht-Monotonic Functions of Higher Order. J. Math. Anal. Appl. **14**, 320–325 (1966).
3. Semigroups of conformal mappings. Seminar on analytic functions, vol. 1. Princeton: Institute for Advanced Study 1957.

Loomis, L.
1. The converse of the Fatou theorem for positive harmonic functions. Trans. Amer. Math. Soc. **53**, 239–250 (1943).

Löwner, K.
1. Über monotone Matrixfunktionen. Math. Z. **38**, 177–216 (1934).
2. Untersuchungen über schlichte konforme Abbildungen des Einheitskreises I. Math. Ann. **89**, 103–121 (1923).

Luke, Y.
1. The special functions and their approximations. New York: 1969.

Menger, K.
1. New foundations of Euclidean geometry. Amer. J. Math. **53**, 721–745 (1931).

Meschkowski, H.
1. Hilbertsche Raume mit Kernfunktion. Berlin: 1962.

Nevanlinna, R.
1. Asymptotische Entwicklungen beschränkter Funktionen und das Stieltjessche Momentproblem. Ann. Acad. Sci. Fenn. A 18–5 (1922).
2. Über beschränkte Funktionen, die in gegebenen Punkten vorgeschriebene Werte annehmen. Ann. Acad. Sci. Fenn./Ser. B 13 (1919).

Nörlund, N.
1. Vorlesungen über Differenzenrechnung. Berlin: 1924.

Perron, O.
1. Die Lehre von Kettenbrüchen. Berlin: 1929.

Pfluger, A.
1. Lectures on conformal mapping. Bloomington: 1969.

Pick, G.
1. Über beschränkte Funktionen mit vorgegebenen Wertzuordnungen. Ann. Acad. Sci. Fenn./Ser. B 15 (1920).
2. Über die Beschränkungen analytischer Funktionen, welche durch vorgegebene Funktionswerte bewirkt werden. Math. Ann. **77**, 7–23 (1916).

Riesz, F., Nagy, B.
1. Leçons d'analyse fonctionelle. Budapest: 1952.

Riesz, M.
1. Sur le probleme des moments. Troisieme Note. Ark. Mat. 17–16 (1923).

Schatten, R.
1. Norm ideals of completely continuous operators. Berlin: 1960.

Schoenberg, I.
1. Metric spaces and positive definite functions. Trans. Amer. Math. Soc. **44**, 522–536 (1938).
2. Remarks to Maurice Fréchets article "Sur la definition...". Ann. of Math. **36**, 724–732 (1935).

Schur, I.
1. Bemerkungen zur Theorie der beschränkten Bilinearformen. J. Reine Angew. Math. **140**, 1–29 (1911).

Seidel, W.
1. Über die Randzuordnung bei konformen Abbildungen. Math. Ann. **104**, 182–243 (1931).

Shohat, J., Tamarkin, J.
1. The problem of moments. New York: 1943.

Stieltjes, T.
1. Recherches sur les fractions continuous. Ann. Fac. Sci. Univ. Toulouse **8**, 1–122 (1894).

Stone, M.
1. Linear transformations in Hilbert space. New York: 1932.

Sz.-Nagy, B.
1. Remarks to the preceeding paper of A. Koranyi. Acta Sci. Math. **17**, 71–75 (1956).

Wall, H.
1. Continued fractions. New York: 1948.

Walsh, J.
1. Interpolation and approximation by rational functions in the complex domain. New York: 1952.

Weyl, H.
1. Über das Pick-Nevanlinnasche Interpolationsproblem und sein infinitesimales Analogon. Ann. of Math. **36**, 230–254 (1935).
2. Über die asymptotische Verteilung der Eigenwerte. Gott. Nachr. p. 110–117, 1911.
3. Über gewöhnliche Differentialgleichungen mit Singularitäten. Math. Ann. **68**, 220–269 (1910).

Widder, D.
1. The Laplace transform. Princeton: 1946.

Wiener, N.
1. The Fourier integral and certain of its applications. Cambridge: 1933.

Wigner, E., von Neumann, J.
1. Significance of Löwners theorem in the quantum theory of collisions. Ann. of Math. **59**, 418–433 (1954).

Index

absolute continuity 53
almost positive matrix 134

Bendat and Sherman 85
Bergman kernel function 92, 141
Bernstein, S. N. 13
Big Bernstein theorem 14

Carathéodory, C. 157
Cauchy interpolation problem 100
Choquet, G. 170
completely monotone function 13

degree of rational function 1
differential 79
divided difference 1

Eberlein, W. 171
entropy 148
exceptional solution 107
exponential representation 27
extended Loewner matrix 40

Fatou theorem 44
Fehrman, J. 173
FitzGerald, C. 140, 144, 151, 167

Gamma function 29
Gram's matrix 8

Hamburger Moment Problem 85
harmonic measure 159
Hilbert-Schmidt norm 66, 78
Hindmarsh, A. 36

infinitely divisible 135
integral representation 20, 21

kernel function 89
Koranyi and Nagy 85, 94

Leibnitz rule 6
linear problem 101
Little Bernstein theorem 13
local property 82, 83, 131
Loewner matrix 39
Loewner-FitzGerald theorem 146
Loewner's theorem 85, 96, 126

matrix function 67
monotone matrix function 68
multiplicity function 58

Newtonian interpolation polynomials 3

one-dimensional perturbations 63
one-dimensional projections 9, 63
operator function 67

Padé table 172
Pick function 18
Pick, G. 94
Pick matrix 34
Pick-Nevanlinna interpolation 172
Pick's theorem 94, 117
positive matrix 7, 88

Radon-Nikodym derivative 54
regularization 11
Remainder theorem 7
reproducing kernel 89
reproducing kernel space 88, 89
resolvent 51
resolvent equation 51, 53
Robertson, A. 170

Schur exponential 10, 93
Schur, I. 9
Schur product 9
Schwarz inequality 8
Seidel, W. 174

self-adjoint 50
sesqui-analytic function 92
spectral theorem 50, 62

univalent 146

Die Grundlehren der mathematischen Wissenschaften in Einzeldarstellungen mit besonderer Berücksichtigung der Anwendungsgebiete

Eine Auswahl

---

23. Pasch: Vorlesungen über neuere Geometrie
41. Steinitz: Vorlesungen über die Theorie der Polyeder
45. Alexandroff: Topologie. Band 1
46. Nevanlinna: Eindeutige analytische Funktionen
63. Eichler: Quadratische Formen und orthogonale Gruppen
102. Nevanlinna/Nevanlinna: Absolute Analysis
114. Mac Lane: Homology
127. Hermes: Enumerability, Decidability, Computability
131. Hirzebruch: Topological Methods in Algebraic Geometry
135. Handbook for Automatic Computation. Vol. 1/Part a: Rutishauser: Description of ALGOL 60
136. Greub: Multilinear Algebra
137. Handbook for Automatic Computation. Vol. 1/Part b: Grau/Hill/Langmaack: Translation of ALGOL 60
138. Hahn: Stability of Motion
139. Mathematische Hilfsmittel des Ingenieurs. 1. Teil
140. Mathematische Hilfsmittel des Ingenieurs. 2. Teil
141. Mathematische Hilfsmittel des Ingenieurs. 3. Teil
142. Mathematische Hilfsmittel des Ingenieurs. 4. Teil
143. Schur/Grunsky: Vorlesungen über Invariantentheorie
144. Weil: Basic Number Theory
145. Butzer/Berens: Semi-Groups of Operators and Approximation
146. Treves: Locally Convex Spaces and Linear Partial Differential Equations
147. Lamotke: Semisimpliziale algebraische Topologie
148. Chandrasekharan: Introduction to Analytic Number Theory
149. Sario/Oikawa: Capacity Functions
150. Iosifescu/Theodorescu: Random Processes and Learning
151. Mandl: Analytical Treatment of One-dimensional Markov Processes
152. Hewitt/Ross: Abstract Harmonic Analysis. Vol. 2: Structure and Analysis for Compact Groups. Analysis on Locally Compact Abelian Groups
153. Federer: Geometric Measure Theory
154. Singer: Bases in Banach Spaces I
155. Müller: Foundations of the Mathematical Theory of Electromagnetic Waves
156. van der Waerden: Mathematical Statistics
157. Prohorov/Rozanov: Probability Theory
159. Köthe: Topological Vector Spaces
160. Agrest/Maksimov: Theory of Incomplete Cylindrical Functions and their Applications
161. Bhatia/Shegö: Stability Theory of Dynamical Systems
162. Nevanlinna: Analytic Functions
163. Stoer/Witzgall: Convexity and Optimization in Finite Dimensions
164. Sario/Nakai: Classification Theory of Riemann Surfaces
165. Mitrinović/Vasić: Analytic Inequalities
166. Grothendieck/Dieudonné: Eléments de Géométrie Algébrique I

167. Chandrasekharan: Arithmetical Functions
168. Palamodov: Linear Differential Operators with Constant Coefficients
170. Lions: Optimal Control Systems Governed by Partial Differential Equations
171. Singer: Best Approximation in Normed Linear Spaces by Elements of Linear Subspaces
172. Bühlmann: Mathematical Methods in Risk Theory
173. F. Maeda/S. Maeda: Theory of Symmetric Lattices
174. Stiefel/Scheifele: Linear and Regular Celestial Mechanics. Perturbed Two-body Motion—Numerical Methods—Canonical Theory
175. Larsen: An Introduction of the Theory of Multipliers
176. Grauert/Remmert: Analytische Stellenalgebren
177. Flügge: Practical Quantum Mechanics I
178. Flügge: Practical Quantum Mechanics II
179. Giraud: Cohomologie non abélienne
180. Landkoff: Foundations of Modern Potential Theory
181. Lions/Magenes: Non-Homogeneous Boundary Value Problems and Applications I
182. Lions/Magenes: Non-Homogeneous Boundary Value Problems and Applications II
183. Lions/Magenes: Non-Homogeneous Boundary Value Problems and Applications III
184. Rosenblatt: Markov Processes. Structure and Asymptotic Behavior
185. Rubinowicz: Sommerfeldsche Polynommethode
186. Wilkinson/Reinsch: Handbook for Automatic Computation II. Linear Algebra
187. Siegel/Moser: Lectures on Celestial Mechanics
188. Warner: Harmonic Analysis on Semi-Simple Lie Groups I
189. Warner: Harmonic Analysis on Semi-Simple Lie Groups II
190. Faith: Algebra: Rings, Modules, and Categories I
192. Mal'cev: Algebraic Systems
193. Pólya/Szegö: Problems and Theorems in Analysis. Vol. 1
194. Igusa: Theta Functions
195. Berberian: Baer *-Rings
196. Athreya/Ney: Branching Processes
197. Benz: Vorlesungen über Geometrie der Algebren
198. Gaal: Linear Analysis and Representation Theory
200. Dold: Lectures on Algebraic Topology
203. Schoeneberg: Elliptic Modular Functions
206. André: Homologie des algèbres commutatives
207. Donoghue: Monotone Matrix Functions and Analytic Continuation
209. Ringel: Map Color Theorem